IMMOBILIZED ENZYMES

PREPARATION AND ENGINEERING TECHNIQUES

IMMOBILIZED ENZYMES
Preparation and
Engineering Techniques

Sidney J. Gutcho

NOYES DATA CORPORATION
Park Ridge, New Jersey   London, England
1974

Published in the United States of America by
Noyes Data Corporation
Noyes Building, Park Ridge, New Jersey 07656

FOREWORD

The detailed, descriptive information in this book is based on recent U.S. patents relating to the immobilization of enzymes.

This book serves a double purpose in that it supplies detailed technical information and can be used as a guide to the U.S. patent literature in this field. By indicating all the information that is significant, and eliminating legal jargon and juristic phraseology, this book presents an advanced, technically oriented review of modern methods of enzyme engineering.

The U.S. patent literature is the largest and most comprehensive collection of technical information in the world. There is more practical, commercial, timely process information assembled here than is available from any other source. The technical information obtained from a patent is extremely reliable and comprehensive; sufficient information must be included to avoid rejection for "insufficient disclosure." These patents include practically all of those issued on the subject in the United States during the period under review, there has been no bias in the selection of patents for inclusion.

The patent literature covers a substantial amount of information not available in the journal literature. The patent literature is a prime source of basic commercially useful information. This information is overlooked by those who rely primarily on the periodical journal literature. It is realized that there is a lag between a patent application on a new process development and the granting of a patent, but it is felt that this may roughly parallel or even anticipate the lag in putting that development into commercial practice.

Many of these patents are being utilized commercially. Whether used or not, they offer opportunities for technological transfer. Also, a major purpose of this book is to describe the number of technical possibilities available, which may open up profitable areas of research and development. The information contained in this book will allow you to establish a sound background before launching into research in this field.

Advanced composition and production methods developed by Noyes Data are employed to bring our new durably bound books to you in a minimum of time. Special techniques are used to close the gap between "manuscript" and "completed book." Industrial technology is progressing so rapidly that time-honored, conventional typesetting, binding and shipping methods are no longer suitable. We have bypassed the delays in the conventional book publishing cycle and provide the user with an effective and convenient means of reviewing up-to-date information in depth.

The Table of Contents is organized in such a way as to serve as a subject index. Other indexes by company, inventor and patent number help in providing easy access to the information contained in this book.

v

15 Reasons Why the U.S. Patent Office Literature Is Important to You —

1. The U.S. patent literature is the largest and most comprehensive collection of technical information in the world. There is more practical commercial process information assembled here than is available from any other source.

2. The technical information obtained from the patent literature is extremely comprehensive; sufficient information must be included to avoid rejection for "insufficient disclosure."

3. The patent literature is a prime source of basic commercially utilizable information. This information is overlooked by those who rely primarily on the periodical journal literature.

4. An important feature of the patent literature is that it can serve to avoid duplication of research and development.

5. Patents, unlike periodical literature, are bound by definition to contain new information, data and ideas.

6. It can serve as a source of new ideas in a different but related field, and may be outside the patent protection offered the original invention.

7. Since claims are narrowly defined, much valuable information is included that may be outside the legal protection afforded by the claims.

8. Patents discuss the difficulties associated with previous research, development or production techniques, and offer a specific method of overcoming problems. This gives clues to current process information that has not been published in periodicals or books.

9. Can aid in process design by providing a selection of alternate techniques. A powerful research and engineering tool.

10. Obtain licenses — many U.S. chemical patents have not been developed commercially.

11. Patents provide an excellent starting point for the next investigator.

12. Frequently, innovations derived from research are first disclosed in the patent literature, prior to coverage in the periodical literature.

13. Patents offer a most valuable method of keeping abreast of latest technologies, serving an individual's own "current awareness" program.

14. Copies of U.S. patents are easily obtained from the U.S. Patent Office at 50¢ a copy.

15. It is a creative source of ideas for those with imagination.

CONTENTS AND SUBJECT INDEX

INTRODUCTION

The investigation of biological processes in plant and animal tissues and in micro-organisms has included the isolation of a large number of enzymes. These en-zymes have been characterized in terms of their structure and the exact mecha-nism of the reactions they control. Knowledge about the role of enzymes and their commercial availability, as pure and abundant reagents, has resulted in their use in industrial processes. However, utilization of enzymes in the production of chemicals has been restricted at times because of the high cost involved in the one-time usage of enzymes as chemical reagents. This major deterrent to the use of enzymes can now be largely eliminated by the use of immobilized enzymes.

Immobilization of enzymes can be achieved by chemical and also by nonchemi-cal or physical procedures. Immobilized enzymes are for the most part insolu-bilized as a result of chemical treatment but they can also be in soluble forms retained by a semipermeable barrier thus becoming in effect immobilized. Re-gardless of the manner of immobilization, immobilized enzymes possess these noteworthy characteristics: they can be separated from the reaction mixture containing the product and any residual reactants and they can be used again as active enzymes in subsequent conversions. Significant and favorable consid-erations in the application of immobilized enzymes also include greater stability for the enzyme, enzyme activity over broad ranges of pH and temperature, elim-ination of enzyme contamination of waste streams, development of continuous processes, and adaptability of immobilized enzymes to a variety of configura-tions and to specific processes.

Four major types of immobilized enzymes have been described: adsorption of an enzyme by a particle surface, entrapment within a gel or by microencapsula-tion, covalent binding of the enzyme to a solid support, and cross-linking to form a solid support consisting of intermolecular cross-links between the reagent and the enzyme. Variations of these four categories have included adsorption followed by cross-linking on the particle surface, adsorption to ion exchange supports, and copolymerization of the enzyme with a polymer-forming material.

1

At the present time, there are three major industrial applications of immobilized enzymes. Columns of DEAE-Sephadex with immobilized aminoacylase are being used for the continuous resolution of DL-amino acids to produce the biologically active L-forms. One such column is currently producing L-methionine at a rate in excess of 1,500 pounds per day. Penicillin amidases covalently bound to derivatized cellulose are being used in the production of penicillin. Immobilized glucoamylase is being used to convert starch to glucose. The commercial potential for this immobilized enzyme is high in view of the large quantity of the soluble form of the enzyme currently utilized in the conversion of starch.

Pharmaceutical and analytical applications have also been demonstrated for immobilized enzymes. It should also be noted that the techniques developed for the immobilization of enzymes are also applicable to the immobilization of non-enzymic proteins, antibodies and other binding proteins, hormones, and many other types of compounds. Immobilization has resulted in the development of the technique of affinity chromatography with a potential for application in a variety of macromolecular-ligand binding systems.

During the past several years, a considerable effort has been expended in the production of enzymes, particularly in the development of microbiological sources with potentials for high yields of enzyme. These enzymes have been isolated from cells by a variety of procedures and purified to show high concentrations of protein and high specific activities. Immobilization with the attractive opportunity for multiple usage of the same enzyme has made enzymology a feasible approach for the production of chemicals and has introduced a new specialty, enzyme engineering, to process development. Microbiology, biochemistry and enzyme engineering represent a combined effort for maximal usage of a vast number of enzymes.

Immobilized enzymes are highly functional biological catalysts. Their application to industrial processes is feasible. Immobilization, as demonstrated thus far for enzymes, can be applied as well to many other biological and immunological procedures.

COVALENT ATTACHMENT

ACTIVATED POLYSACCHARIDES AS SUPPORTS

Diazotized Anthranilates of Starch

Carriers in use for enzyme insolubilization are synthetic copolymers based on acrylamide containing aromatic amine residues which are diazotized for coupling to enzymes, polyacrylamide gel which entraps the enzyme and immobilizes it, and polyionic carriers such as diethylaminoethylcellulose which physically adsorbs the enzyme. Recently chemically coupled enzymes have also been prepared through diazotized aminoaarylsilane substituted porous glass. Such water-insoluble enzymes are used either in suspension or in packed columns.

C.L. Mehltretter and F.B. Weakley; U.S. Patent 3,745,088; July 10, 1973; assigned to the U.S. Secretary of Agriculture have reported a process for preparing active water-insoluble enzymes. Starch anthranilates having a degree of substitution (DS) of anthranilate substituents of from about 0.01 to 0.10 are diazotized by reacting the starch anthranilates with nitrous acid at a pH of from 1 to 3 and at a temperature of 0° to 10°C. The diazotized derivatives are then reacted at a pH of about 8 to 9 with an excess of enzyme having substituents capable of coupling with diazonium groups. The water-insoluble enzyme is then isolated and activated.

The preparation of a water-insoluble enzyme having the general structure

$$St-O-\overset{\overset{\displaystyle O}{\|}}{C}$$

with —N=N—E

where St = starch and E = enzyme is as described.

3

The preferred enzyme is papain, but many other enzymes (e.g., glucoamylases and isomerases) can be insolubilized and stabilized by the method. A large excess of enzyme is preferred to avoid coupling the active sites in the enzymes to too great an extent and possibly inactivating the enzyme. It is believed that coupling of the diazonium chloride group of the starch anthranilate to a portion of the tyrosine moieties in the enzyme occurs in such a manner that the enzyme's active sites are not greatly affected.

The water-insoluble enzymes prepared in this manner can be recovered after each use by centrifuging or filtering and recycled. The moist product is stored in a refrigerator at about 10°C until it is used. Before each use the enzyme can be activated by any conventional method. Hydrogen sulfide was used in the examples only because it did not interfere with subsequent analyses as would cysteine in combination with the sodium salt of ethylenediaminetetraacetic acid. The following examples illustrate the process.

Example 1: Preparation of Starch Anthranilate of DS 0.10 — To 50 grams pearl cornstarch (44 grams, oven-dry basis) slurried in 75 ml of water by mechanical agitation was added 5 grams of sodium carbonate followed by 5 grams of isatoic anhydride. The mixture was stirred for 4 hours at about 45°C and the final pH of the reaction mixture was 9.2. The starch anthranilate was filtered by suction and washed three times by agitation in 60 ml of water with intermediate filtration. The last wash was adjusted to pH 5 with dilute hydrochloric acid. The product was air-dried and weighed 46 grams (oven-dry basis). Nitrogen analysis indicated the DS to be 0.10.

Diazotization of Starch Anthranilate of DS 0.10 — 5 grams of starch anthranilate of DS 0.10 was mechanically stirred in 60 ml of water and adjusted to pH 2 with 4 N hydrochloric acid. The mixture was cooled to 0°C and 0.8 gram of sodium nitrite added and the whole stirred at 0° to 5°C for 10 minutes with the addition of more HCl to maintain the pH at about 2. The mixture then stood at about 5°C for an hour with occasional stirring. The diazotized starch anthranilate was then filtered by suction and washed on the filter with water several times to remove nitrous acid (see U.S. Patent 3,499,886).

Coupling to Papain — The wet diazonium chloride prepared from starch anthranilate was added to a solution of 20 grams of technical grade papain in 600 ml of water and stirred at pH 8.4 for 30 minutes and allowed to stand at 10°C overnight. The water-insoluble coupled enzyme was separated by filtration and washed four times with 100 ml of water. The moist product was stored at 10°C until assayed for enzyme activity. The final wash water was free of active papain.

The water-insoluble papain prepared as described above was assayed daily for 10 days and isolated each day for reuse. The procedure was as follows: Wet insolubilized papain (8.2 grams, 66% water) was activated in 100 ml of half-saturated H_2S-H_2O for 1 hour with occasional stirring, then sparged with nitrogen, and was isolated by centrifugation. The activated insoluble papain was admixed with 10 ml of 6% Hammarsten casein and immediately adjusted to pH 5.1 with citrate buffer. The system was incubated for 3 hours at 40°C with frequent stirring and then diluted to 100 grams with water and adjusted to pH

4.2 by dropwise addition of 4 N HCl. The diluted digest was centrifuged and 1 ml of supernatant removed for colorimetric assay with ninhydrin reagent solution. Difference between the ninhydrin color yield of the digested casein and that of an undigested casein was used for comparison with the color yields obtained when varying amounts of the soluble papain were used in casein hydrolyses. By this assay, 1 gram (oven-dry basis) of the water-insoluble papain was found, after several leachings with water, to reach a relatively constant level of proteolytic activity equivalent to approximately 20 mg soluble papain. The water-insoluble papain was isolated from the digest for reuse by exhaustive aqueous extraction at pH 8 to remove substrate. Thereafter, the insolubilized papain, following reactivation, was reused in the hydrolysis of casein.

The table below shows the activity of the chemically coupled papain after 10 successive hydrolysis runs using a casein substrate.

Hydrolysis Runs	Sample Activity, mg/g
1	51
2	53
3	52
4	23
5	20
6	20
7	20
8	21
9	21
10	20

A nearly constant level of proteolytic activity was obtained after three successive hydrolyses of casein.

Example 2: Preparation of Cyanoethylstarch Anthranilate – 44 grams of cyanoethylstarch of DS 1.0 was reacted with isatoic anhydride as described in Example 1 to yield 43.5 grams of cyanoethylstarch anthranilate of DS 0.01.

Diazotization of Cyanoethylstarch Anthranilate – 5 grams of the cyanoethylstarch anthranilate was diazotized as described in Example 1 to yield the wet diazonium chloride.

Coupling to Papain – The wet diazonium chloride was stirred for 30 minutes at pH 8.5 in a solution of 20 grams of technical grade papain in 500 ml of water. The reaction mixture then stood at 10°C overnight and the water-insoluble papain was recovered by centrifugation and washed well with water as in Example 1. The final wash water was devoid of active papain.

Activity of the product was determined as described in Example 1. A constant level of proteolytic activity equivalent to 140 mg of soluble papain per gram of dry product was obtained after three successive hydrolyses of casein.

Diazotized Anthranilate Ester of Polyanhydroglucose

N.E. Franks; U.S. Patent 3,647,630; March 7, 1972; assigned to The Proctor & Gamble Company has developed an enzymatically active, water-soluble composition consisting essentially of a protease enzyme covalently bonded to a water-insoluble matrix which is a diazotized anthranilate ester of a polyanhydroglucose compound. The composition is useful for applications where it is desired to employ a water-insoluble enzymatically active catalyst having a preformed matrix with good mechanical characteristics and specifically selected enzyme activity.

The matrix is a material having a formula

$$\text{polyanhydroglucose group-}O-\overset{\overset{\textstyle O}{\|}}{C}-C\overset{\displaystyle N=N^{\oplus}X^{\ominus}}{\underset{\displaystyle C}{\overset{\displaystyle C}{\diagup}}\;\;\;\;\overset{\displaystyle C}{\underset{\displaystyle C}{\|}}}$$

in which the polyanhydroglucose group is cellulose, starch, or dextran cross-linked with epichlorohydrin; X is a chlorine or bromine anion; the matrix containing from 0.5 to 1.5% nitrogen.

The water-insoluble matrix is prepared by reacting a polymeric polyhydroxylic material which is a polyanhydroglucose compound with isatoic anhydride to prepare a reaction product which is an anthranilate ester of the polyanhydroglucose compound. The anthranilate ester is then diazotized to form the water-insoluble matrix which is then reacted with a protease enzyme to form the insolubilized enzymatically-active composition.

The starting polyanhydroglucose compound can be cellulose, starch, or an epichlorohydrin cross-linked dextran. Each of these materials should be used in their insoluble form. This includes particulate forms including fibrous materials and gels. The cellulose can have a molecular weight in the range of 20,000 to 2,000,000 and preferably be in the range of 200,000 to 2,000,000. Suitable celluloses are those obtained from seed fibers such as cotton, kapok, and other flosses, woody fibers such as pine, bast fibers such as straw, flax, hemp, ramie, and leaf fibers such as sisal and manila hemp. Cellulose, a carbohydrate constituent of the walls and skeletons of vegetable cells, is a polymer of the glucose residue units:

$$\left[\begin{array}{c} \text{structure of glucose residue units} \end{array} \right]$$

The starting water-insoluble polyanhydroglucose material, whether it is cellulose, starch, or a cross-linked dextran is reacted with isatoic anhydride to form an anthranilate ester. The reaction of isatoic anhydride with the polyhydroxylic matrix can be described as below:

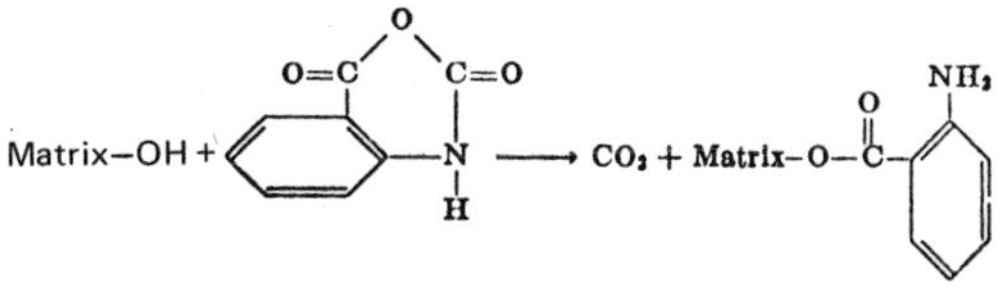

The reaction is best performed in an aqueous environment. The anhydride portion of the isatoic anhydride is in an activated state and readily reacts with nucleophilic species. By performing this reaction in a slightly alkaline environment, some of the hydroxyl residues in the matrix are converted to the alkoxide form (Matrix−O−) which react with the anhydride. Aqueous sodium carbonate achieves this end; aqueous sodium bicarbonate also suffices. Not all of the isatoic anhydride reacts with the matrix, but is hydrolyzed instead to sodium anthranilate and CO_2.

A reaction temperature of 2° to 50°C is the desirable range in which to perform the reaction. Stirring is necessary to assure the uniform reaction of isatoic anhydride with the matrix. Isatoic anhydride has a solubility in water at room temperature of less than 0.7% (wt/vol). To achieve a homogeneous reaction, it is desirable to add an inert cosolvent to the reaction mixture.

Dimethylformamide, N-methylpyrrolidine, or dimethylacetamide added to the reaction mixture in equal parts, or less, dissolve the isatoic anhydride in the reaction mixture. Reaction times of 2 hours are sufficient, but reaction times of up to 24 hours are acceptable. The weight ratio of the matrix to isatoic anhydride can be as great as 1:1. In practice, it is best held at 5 to 10:1. This ratio yields a matrix with a N_2 content of 0.55 to 1.1% and a degree of substitution of 0.1 to 0.2. The ester formed by the preceding reaction has the following formula:

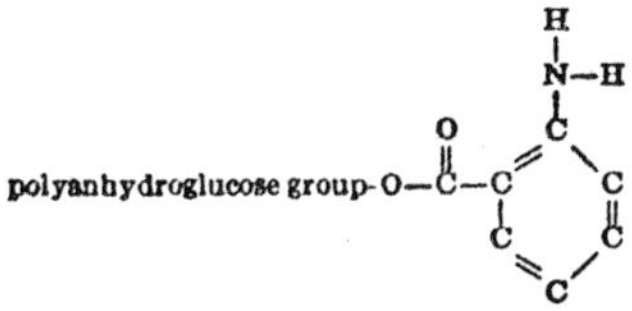

The matrix anthranilate ester is separated from the reaction mixture by either filtration or centrifugation. Removal of unreacted isatoic anhydride from the matrix is best accomplished by washing the matrix with several volumes of the inert cosolvent employed in the reaction. It is also necessary to remove the isatoic anhydride hydrolysis products, anthranilic acid and sodium anthranilate. This is best accomplished by exhaustive washing with aqueous bicarbonate or carbonate. This is easily monitored by the cessation of fluorescence of the washings under ultraviolet light. The next step in the preparation of the matrix is the diazotization of the anthranilate ester of the polyanhydroglucose material. This is a known reaction and involves a reaction between nitrous acid HNO_2 and the ester. The reaction product from the diazotization step is a compound having the formula given above. The X anion will be determined by whatever means is employed in the diazotization step; typically, it is a halogen or a weak organic acid. In a classic embodiment in which nitrous acid is generated in situ by a reaction between acetic acid or propionic acid and sodium nitrite, the X anion is an acetoxy group or a propoxy group. Hydrochloric acid and hydrobromic acid reacted with sodium nitrite result in the anion being chloride and bromide respectively.

In order to obtain the maximum benefits, the insolubilized composition should contain from 0.5% to 1.5%, preferably from 0.75% to 1.3% nitrogen. When less than 0.5% is present, an insufficient amount of enzyme attaching sites are available. No additional need appears to exist for exceeding 1.5% nitrogen.

Example 1: (A) Preparation of Anthranilate Ester of Epichlorohydrin-Cross-Linked Dextran — A cross-linked dextran (5 grams) (Sephadex G-25) having a molecular weight exclusion limit approximate of 5,000 and a particle size of 50 to 150 microns was suspended in 25 ml of 0.1 M sodium bicarbonate and allowed to equilibrate for 30 minutes. This mixture was stirred at room temperature and a solution of 1 gram of isatoic anhydride in 5 ml of dimethylformamide was added to the mixture. An additional 5 ml of dimethylformamide was added to the mixture to bring about solution of the isatoic anhydride. The mixture was stirred for two hours at room temperature during which period the anthranilate ester of cross-linked dextran was formed as the desired reaction product.

Work-up of the anthranilate ester of dextran was performed by filtering and washing with dimethylformamide. Sodium anthranilate, arising from hydrolysis of isatoic anhydride during the reaction, was strongly absorbed by the ester reaction product. To insure complete removal of sodium anthranilate, it was necessary to wash the ester reaction product with saturated aqueous sodium bicarbonate until the washings no longer fluoresced under UV light. The ester reaction product was then washed with distilled water until the washings were neutral to pH indicator paper. The cross-linked dextran anthranilate ester which had a nitrogen content of 1.2% was stored in moist form until needed.

(B) Preparation of Diazotized Cross-Linked Dextran Anthranilate — Moist cross-linked dextran anthranilate (5 grams) from Part A above, was suspended in 23 milliliters of distilled water. Glacial acetic acid (2 ml) was added to this mixture with stirring. The mixture was chilled to 0° to 2°C, and a concentrated solution of sodium nitrite in water was added in dropwise with stirring. Addition of sodium nitrite was halted when an excess of this reagent was present as gauged by starch-potassium iodide paper. A diazotization reaction occurred which formed a diazotized dextran anthranilate matrix. This reaction product was stirred an additional 30 minutes before it was isolated by filtration. The diazotized matrix was washed with 200 ml of ice-water to remove excess acetic acid and sodium nitrite.

(C) Attaching Protease Enzyme to the Diazotized Anthranilate Ester of Cross-Linked Dextran — Crystalline Alcalase (100 mg) was dissolved in 5 ml of 0.2 M sodium acetate:0.02 M calcium chloride buffer. The pH of this solution was ad-

justed to 8 to 8.5 with N sodium hydroxide. This solution was cooled to 0° to 2°C, and the diazotized cross-linked dextran anthranilate matrix prepared in Part B above was added with stirring. Sodium hydroxide (a one normal solution) was added dropwise to maintain the pH between 8 and 8.5. After two hours, the mixture was allowed to come to room temperature. After base uptake ceased, a column was poured and it was washed with 0.1 M borate, pH 8.8. Unreacted diazo groups on the matrix decomposed during this washing to generate N_2; the matrix was transferred to a flask containing 0.2 M borate buffer, pH 9.0. After evolution of N_2 had ceased, the reddish matrix was washed with more 0.1 M borate and the activity of the matrix having the insolubilized protease enzyme attached to the diazotized anthranilate ester of cross-linked dextran was measured by a column technique.

The moist active matrix was dried in vacuo at 100°C and found to contain 32% solids; hydrolysis of the dried matrix with 6 N hydrochloric acid and quantitative amino acid analysis of the hydrolyzate showed the presence of 3 to 4 mg of enzyme/gram of the dried matrix.

Example 2: In this example a protease enzyme, Bacterial Proteinase Novo, was covalently bonded to the diazotized epichlorohydrin-cross-linked dextran anthranilate matrix prepared in Parts A and B of Example 1.

The matrix was added to a solution of 100 mg of crystalline Bacterial Proteinase Novo in 5 ml, 0.2 M sodium acetate, 0.02 calcium chloride at pH 8 to 8.5. The coupling was initiated at 0°C and the pH was maintained at 7.5 to 8.5 by addition of N sodium hydroxide to the stirred mixture. After 3 hours, a column was poured with the insolubilized enzyme-matrix adduct; the column was washed with 0.1 M borate buffer, pH 8.8. The enzyme-matrix adduct was stored at 5°C in a 0.13 M sodium chloride, 0.03 M phosphate buffer, pH 7.0. A portion of the moist matrix was dried in vacuo at 100°C to yield 27.5% solids. Hydrolysis of the dried matrix in 6 N HCl followed by quantitative amino acid analysis of the hydrolyzate showed the presence of 2 to 3 mg enzyme/g of dry matrix.

Alkylenediamine Binding of Dialdehyde Starch

E. Katchalski, L. Goldstein, Y. Levin and S. Blumberg; U.S. Patent 3,706,633; December 19, 1972; assigned to Monsanto Company has prepared water-insoluble, enzymatically active compositions containing dialdehyde starch having from about 15 to about 100% of 2,3-alcohol groups thereof oxidized to dialdehyde groups, an active enzyme, and an alkylenediamine binding the enzyme to the starch. These compositions are prepared by condensing the dialdehyde starch with an alkylenediamine to produce a polymeric product, reducing and diazotizing the product, so as to produce a polymeric polydiazonium salt which is then coupled with an active enzyme. A water-insoluble, active enzyme modified dialdehyde starch conjugate containing up to about 10% by weight, or more, of the enzyme can be prepared.

In preparing the compositions, the dialdehyde starch is first condensed, at about room temperature and in a suitable reaction medium, with the alkylenediamine to produce a highly cross-linked polymeric product which is then reduced with a complex metal hydride such as lithium aluminum hydride or an alkali metal borohydride such as sodium borohydride, lithium borohydride or the like. The product is then diazotized to produce a corresponding polydiazonium salt which, in turn, is coupled with the desired enzyme. The preparation of the enzyme carrier resin is believed to proceed according to the following scheme:

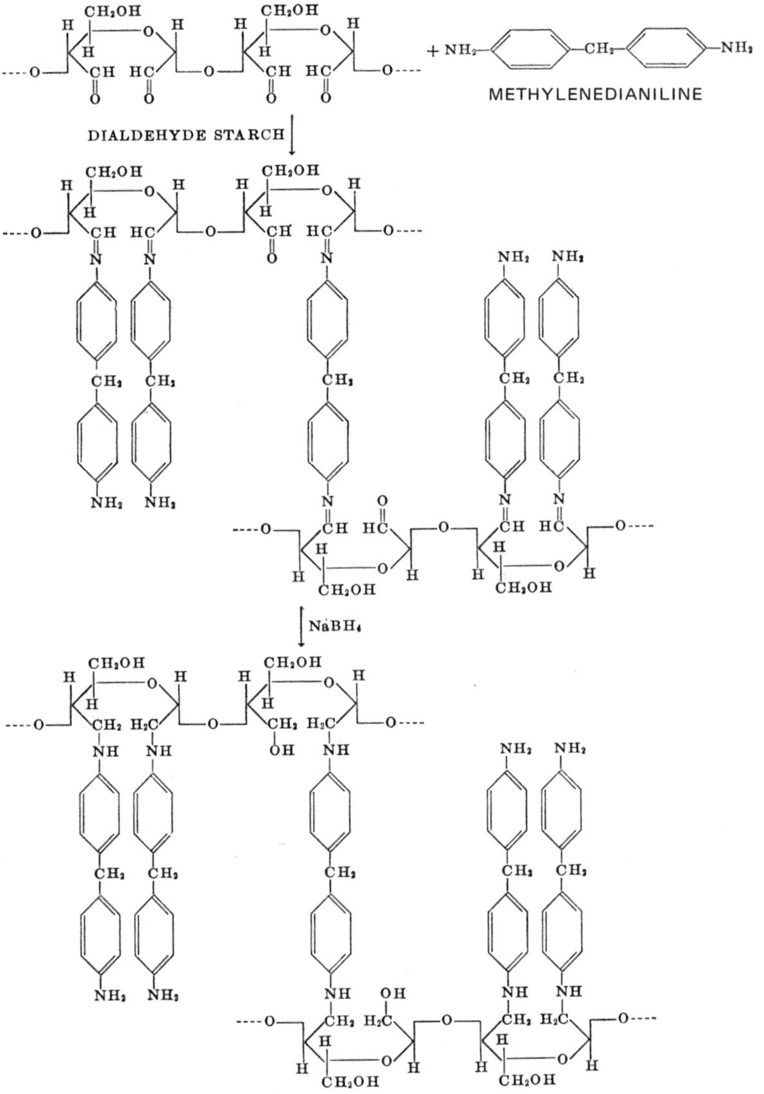

Example 1: Preparation of Carrier Resin with Methylenedianiline – 90% oxidized dialdehyde starch (10 grams) is suspended in water (200 ml) and stirred for about 10 to 15 min at room temperature to produce a fine slurry, 2 molar

carbonate buffer, pH 10.5 (40 ml) is added thereto and the suspension is slowly poured into a vigorously stirred, 10 weight percent solution of methylenedianiline in methanol (300 ml). Thereafter the stirring of the resulting reaction mixture is continued for about 2 to 3 days and then an insoluble polymeric Schiff's base is recovered therefrom, washed with methanol, resuspended in water, and reduced with sodium borohydride (40 grams).

The resulting mixture is then neutralized with acetic acid and the produced resin washed with water and methanol, and thereafter refluxed in methanol (3 to 4 changes, 300 ml each) so as to remove any methanol soluble aromatic amines that may be present, until a negative test is obtained with N,N-dimethylaminobenzaldehyde (Ehrlich's Reagent). Thereafter the resin is filtered and dried. About 14 to 15 grams of dry resin is produced.

The diazotization capacity of the produced resin is estimated by coupling the polydiazonium salt with p-bromophenol and determining the nitrogen and bromine content of the resulting product. The resin (50 mg) is suspended in 50 weight percent acetic acid (4 ml) and stirred for one hour over ice. An aqueous solution of sodium nitrate (10 mg per ml of water) is then added dropwise to the chilled suspension. The obtained diazotization mixture is further stirred for one hour over ice and then brought to pH 8.5 by dropwise addition of aqueous 5 N solution of sodium hydroxide.

The produced precipitate is recovered, washed with cold 0.2 M phosphate buffer, pH 7.8, and suspended in the same buffer (5 ml). A solution of p-bromophenol (50 mg dissolved in water by the dropwise addition of aqueous 2 N sodium hydroxide) is then added to the chilled suspension. The resulting reaction mixture is stirred for about 16 hours at about 4°C and the precipitate separated by filtration, washed with 0.1 M carbonate buffer (pH 10.5), water, and methanol, and dried in vacuum over phosphorus pentoxide.

Example 2: Preparation of Carrier Resin with Hexamethylenediamine – 90% oxidized dialdehyde starch (1.6 grams) is suspended in water (20 ml), and the resulting suspension brought to pH 11.2 and vigorously stirred. An aqueous solution of hexamethylenediamine (1.16 grams in 2 ml of water) is slowly added thereto. The reaction is permitted to proceed for about 16 hours at room temperature, the pH of the reaction mixture being monitored by a pH-stat. Thereafter a dark brown precipitate is recovered from the reaction mixture, washed with water, methanol, and diethyl ether, and then dried in vacuum over concentrated sulfuric acid. About 1.5 grams of resin is obtained.

A portion of the obtained resin (200 mg) is suspended in water (5 ml), vigorously stirred, and a methanol solution of methylenedianiline (about 150 mg in 1.5 ml of methanol) is added slowly to the stirred suspension. The pH of the resulting admixture is maintained at 10.8 by the addition of aqueous 0.1 N sodium hydroxide solution as required. The admixture is stirred at room temperature for about 16 hours. Thereafter an insoluble resin is recovered therefrom, washed with methanol until free from methanol-soluble aromatic amines as indicated by a negative test with N,N-dimethylaminobenzaldehyde (Ehrlich's Reagent), washed with diethyl ether and dried in vacuum over sulfuric acid. Net weight of the produced dry resin is 220 mg.

The produced resin (146 mg) is suspended in water (5 ml) and sodium boro-hydride (200 mg) added thereto. The resulting admixture is stirred for about 16 hours at room temperature. The obtained reduced resin is recovered from the admixture, washed with an aqueous acetic solution and then with water.

The washed resin is suspended in cold aqueous 1 N HCl solution (3 ml), aqueous sodium nitrite solution (1.46 mg $NaNO_2$ in 1 ml of water) added dropwise thereto, and the suspension stirred for 90 minutes. Thereafter pH of the suspension is adjusted to 7.7 by the addition of aqueous 0.1 N sodium hydroxide solution and the suspension is employed directly in the coupling of enzyme to the produced diazotized resin.

Example 3: Coupling of Papain to the Carrier Resin — Papain (10 mg) is dissolved in 0.1 M phosphate buffer, pH 7.7, (10 ml) and a portion of the suspension of diazotized resin produced in Example 2 is added to the solution. The resulting admixture is stirred for about 16 hours at 4°C. Thereafter the resin is recovered, washed with water, and suspended in 0.1 M phosphate buffer, pH 6.5.

The obtained insoluble papain bearing resin exhibits enzymatic activity as determined by a titrimetric assay at 25°C, pH 6.3, using benzoyl-L-arginine ethyl ester as substrate and aqueous 0.1 N sodium hydroxide solution as titrant.

Example 4: Coupling of Mercuripapain to Carrier Resin — Dialdehyde starch methylenedianiline resin (100 mg) prepared and diazotized as in Example 1, is washed and the washed polydiazonium salt precipitate is suspended in 0.2 M phosphate buffer, pH 7.8 (6 ml), a solution of mercuripapain (8 to 10 mg) is added dropwise thereto, and the resulting admixture stirred for about 16 hours at 4°C. The produced insoluble mercuripapain-carrier resin conjugate is recovered by filtration on a Buchner funnel, washed with aqueous 1 M potassium chloride solution (200 ml), then with water (100 ml), and resuspended in water.

The produced conjugate is assayed in the same manner as papain and exhibits an enzymatic activity of about 60% of that exhibited by unbound mercuripapain. The mercuripapain-resin conjugate retains enzymatic activity upon storage as an aqueous suspension or as a lyophilized or dried powder.

The caseinolytic activity of the produced conjugate is similar to that of the native enzyme as calculated on the basis of the respective esterase activities. The conjugate can be easily removed from a digestion mixture by filtration or centrifugation, it has good flow properties and thus is suitable for column work.

Example 5: Carrier Resin-Polytyrosyltrypsin Conjugate — Dialdehyde starch-methylenedianiline resin (100 mg) prepared as in Example 1 is suspended in aqueous 50 weight percent acetic acid solution (8 ml) and the suspension stirred for one hour over ice and afterwards an aqueous solution of sodium nitrite (20 milligrams in one milliliter of water) is added dropwise to the chilled suspension. The resulting diazotization mixture is stirred for one hour over ice and then brought to pH 8.5 by the dropwise addition of aqueous 5 N solution of sodium hydroxide. Crushed ice is added as required to maintain the mixture at a relatively low temperature.

The polydiazonium salt of the resin separates as a dark brown, lumpy precipitate which is recovered on a Buchner funnel, washed with cold 0.2 M phosphate buffer (pH 7.8), and suspended in the same buffer (10 ml).

A solution of polytyrosyltrypsin (30 mg in 8 ml of aqueous 0.001 N HCl) is then added to a chilled suspension of the diazotized resin and the coupling reaction allowed to proceed for about 16 hours at 4°C with stirring. The carrier resin-polytyrosyl trypsin conjugate is recovered from the suspension by filtration, washed with an aqueous 1 M potassium chloride solution (100 to 200 ml), and then with water. Thereafter the conjugate is resuspended in water or in 0.1 M phosphate buffer, pH 7.0.

Enzymatic activity of the conjugate is assayed in the same manner as for trypsin at pH 10, the amount of active enzyme being calculated from the rate of substrate hydrolysis. The protease activity of the conjugate is determined at pH 7.6 by the casein digestion method.

Diazophenoxy and Isothiocyanatophenoxyhydroxypropylcelluloses

S.A. Barker, P.J. Somers and R. Epton; U.S. Patent 3,627,638; December 14, 1971; assigned to Ranks Hovis McDougall Limited, England have prepared α-, β- and γ-amylase chemically coupled to p-diazophenoxyhydroxypropylcellulose and α- and β-amylase chemically coupled to p-isothiocyanatophenoxyhydroxypropylcellulose.

Water-insoluble preparations of α-, β- or γ-amylase may be made by chemical reaction at 0° to 5°C of the amylase dissolved in a buffer within a pH range of 6.3 to 7.7 (preferably 7.6 to 7.7) with the p-diazophenoxyhydroxypropyl ether of cellulose. Unreacted diazo groups in the cellulose derivative are annealed by reaction with either β-naphthol or phenol. Preferably microcrystalline cellulose is used for the preparation of this ether and the degree of substitution of ether groups in the cellulose can be 13 to 56.2 microequivalents (preferably 13 microequivalents) of p-diazophenoxyhydroxypropyl ether groups per gram of cellulose. Active water-insoluble preparations of α-amylase, β-amylase, glucamylase (γ-amylase) can be obtained by this process which are more heat stable when suspended in an aqueous buffer (0.02 M) than the corresponding soluble enzyme. Preferably the buffer should have that pH at which the enzyme displays maximum enzyme activity towards its substrate.

Alternatively, water-insoluble preparations of α- or β-amylase may be made by chemical reaction at 0° to 5°C of α- or β-amylase dissolved in a buffer (preferably 0.05 M borate buffer, pH 8.6) with the p-isothiocyanatophenoxyhydroxypropyl ether of cellulose. Preferably microcrystalline cellulose is used for the preparation of this ether and the degree of substitution of ether groups in the cellulose can be 13 to 56.2 microequivalents (preferably 13 microequivalents) of p-isothiocyanatophenoxyhydroxypropyl ether groups per gram of cellulose. Active water-insoluble preparations of α-amylase and β-amylase but not glucamylase (γ-amylase) can be obtained by this method which are more heat stable when suspended in an aqueous buffer (0.02 M) than the corresponding soluble enzyme. Preferably the buffer should have that pH at which the enzyme displays maximum enzymic activity towards its substrate.

The particular merits for providing water-insoluble enzymes are that this process can provide a product with a high retention of activity when calculated as a percentage of the activity which that amount of enzyme protein bound to the cellulose derivative would display in its original soluble form. The second advantage is that the process may be particularly advantageous for exoenzymes such as β- and γ-amylase. The third advantage is that the use of microcrystalline cellulose in the preparation of the ether affords a dense hydrophilic carrier available in a fine particulate form for maximum surface exposure yet easily recoverable after use by centrifugation or filtration. The fourth advantage is the much greater heat stability of the water-insoluble enzymes which may be obtained by the process, compared with the corresponding soluble enzyme, giving a greater shelf life, a greater retention of activity at operating temperatures and enabling maximum repetitive use to be made of the enzyme.

Example 1: p-Aminophenoxyhydroxypropylcellulose hydrochloride (100 mg, 20.6 μeq ether linkage/g) was placed in a stoppered test tube and magnetically stirred into a slurry with phosphate buffer (3.5 M, pH 6.8, 0.5 ml). Thiophosgene solution (10%, 0.2 ml) in carbon tetrachloride was added and stirring continued for 20 minutes when a further aliquot (0.2 ml) of thiophosgene solution was added. After stirring for a further 20 minutes acetone (15 ml) was added and the solid p-isothiocyanatophenoxyhydroxypropylcellulose recovered by centrifugation. The washing cycle was repeated twice with sodium bicarbonate solution (0.5 M, 15 ml) and twice with borate buffer (0.05 M, pH 8.6, 15 ml).

After decantation of the final washings a solution of β-amylase (5 mg) in borate buffer (0.05 M, pH 8.6, 1 ml) was added and coupling allowed to proceed with gentle magnitic stirring for 48 hours at 0° to 5°C. The water-insoluble β-amylase derivative was subjected to five cycles of alternate washing with acetate buffer (0.02 M, pH 4.8, 15 ml) and a solution of sodium chloride (1.0 M, 15 ml) in the same buffer. After two further washings with acetate buffer (15 ml) and final decantation of the washings the β-amylase derivative was resuspended in the same buffer (10 ml).

Example 2: p-Aminophenoxyhydroxypropylcellulose hydrochloride (100 mg, 20.7 μeq ether linkage/g) was stirred magnetically at 0°C with hydrochloric acid (1N, 5 ml). Sodium nitrite solution (2%, 4 ml) precooled to 0°C, was added and stirring continued for 15 minutes. The p-diazophenoxyhydroxypropylcellulose was washed four times with phosphate buffer (0.075 M, pH 7.6 to 7.7, 15 ml) at 0°C. After decantation of the final washings a solution of β-amylase (5 mg) in phosphate buffer (0.075 M, pH 7.6 to 7.7, 1 ml) was added and coupling allowed to proceed with gentle magnetic stirring for 48 hours at 0° to 5°C. A solution of phenol (0.01%, 5 ml) in saturated aqueous sodium acetate at 0°C was then added. After a further 15 minutes stirring the water-insoluble β-amylase derivative was recovered by centrifugation. After discarding the supernatant, the derivative was subjected to washing and suspended in acetate buffer (0.02 M, pH 4.8, 10 ml).

Example 3: p-Diazophenoxyhydroxypropylcellulose was washed four times with phosphate buffer (0.075 M, pH 7.6 to 7.7, 15 ml) at 0°C. After decantation of the final washings a solution of γ-amylase (5 mg protein) in phosphate buffer (0.075 M, pH 7.6 to 7.7, 1 ml) was added and coupling allowed to proceed with gentle magnetic stirring for 48 hours at 0° to 5°C. A solution of phenol (0.01%, 5 ml) in saturated sodium acetate at 0°C was then added. After a further 15 minutes stirring the water-insoluble γ-amylase derivative was recovered by centrifugation. After discarding the supernatant, the derivative was subjected to washing and suspended in acetate buffer (0.02 M, pH 4.5, 10 ml).

Prep. No.	Enzyme	Enzyme Units*/mg Free Protein	Mg Bound Protein/ 100 mg Derivative	Enzyme Units*/mg Bound Protein	% Activity Retained by Enzyme After Coupling
15	Ex 2	499.9	4.24	83.3	16.6
16	Ex 3	28.2	1.25	5.01	17.7

*One amylase unit was taken as that which liberated reducing sugar equivalent to 1 mg maltose at 20°C (β-amylase) or 1 mg glucose at 45°C (γ-amylase) in 3 minutes.

S.A. Barker, P.J. Somers and R. Epton; U.S. Patent 3,702,804; November 14, 1972; assigned to Ranks Hovis McDougall Limited, England have described a process for the preparation of a water-insoluble pullulanase, carboxypeptidase or dextranase which process comprises reacting at 0° to 5°C the pullulanase, carboxypeptidase or dextranase dissolved in a buffer within a pH range of 6.5 to 8.6 with the p-diazophenoxyhydroxypropyl ether of cellulose.

Example 1: A sample (100 mg) of the p-aminophenoxyhydroxypropylcellulose ether hydrochloride was placed in a stoppered test tube together with an aliquot (5 ml) of hydrochloric acid (1N). The tube was placed in an ice bath and magnetically stirred for 15 minutes when 4 ml of 2% ice-cold sodium nitrite was added. After a further 15 minutes the tube was centrifuged, the supernatant discarded, 15 ml aliquots of ice-cold phosphate buffer (0.075M, pH 7.6 to 7.7) added and the contents stirred magnetically for 15 minutes.

The washing cycle was repeated three times at 0°C. The centrifuged cellulose derivative was taken, carboxypeptidase (8.2 mg) in phosphate buffer 0.075M, pH 7.6 to 7.7 (1 ml) was added, and the tube stirred magnetically at 0° to 5°C for 24 hours when ice-cold β-naphthol (5 ml, 0.1%) in saturated sodium acetate was added.

After stirring a further 15 minutes at 0°C, the water-insoluble carboxypeptidase derivative was subjected to five cycles of washing with 15 ml phosphate buffer (0.02M, pH 6.9) and sodium chloride (0.5M) 15 ml in the same buffer. The carboxypeptidase derivative was finally washed three times with phosphate buffer (0.02M, pH 6.9), and phosphate buffer (0.075M, pH 7.6 to 7.7) 10 ml added.

Carboxypeptidase activity was assayed by stirring the enzyme suspension (1 ml) with N-carbobenzoxyglycyl L. phenylalanine solution (0.1%) 10 ml in phosphate buffer (0.075M, pH 7.6 to 7.7) at 30°C. Samples (1 ml) were removed at times zero, 5, 10, 15 minutes and pipetted directly into assay tubes containing ninhydrin reagent (1 ml).

The results were as follows: mg bound protein/100 mg derivative, 1.68; enzyme units/mg bound protein, 6.88; and percent activity retained by enzyme after coupling, 15.6. One carboxypeptidase unit is that which liberates 1 μmol of L-phenylalanine per mg protein in 1 minute at 30°C and pH 7.6 to 7.7.

Example 2: A sample (100 mg) of the p-aminophenoxyhydroxypropylcellulose ether hydrochloride was placed in a stoppered test tube together with an aliquot (5 ml) of hydrochloric acid (1N). The tube was placed in an ice bath and magnetically stirred for 15 minutes when 4 ml of 2% ice-cold sodium nitrite was added. After a further 15 minutes the tube was centrifuged, the supernatant discarded, 1.5 ml of ice-cold phosphate buffer (0.075M, pH 7.6) added and the contents stirred magnetically for 15 minutes. The washing cycle was repeated three times at 0°C. The centrifuged cellulose derivative was taken, dextranase (1 ml) in 0.075M phosphate buffer pH 7.6 was added and the tube stirred magnetically at 0° to 5°C for 24 hours when ice-cold β-naphthol (5 ml, 0.1%) in saturated sodium acetate was added.

After stirring a further 15 minutes at 0°C, the water-insoluble dextranase derivative was subjected to five cycles of washing with 15 ml phosphate buffer (0.2M, pH 6.9) and sodium chloride (0.5M) 15 ml in the same buffer. The insoluble dextranase derivative was finally washed three times with citrate buffer (0.2M, pH 5), the stirrer bars removed and 10 ml 0.2M citrate buffer pH 5 was added.

Bromoacetylcellulose

A. Patchornik; U.S. Patent 3,278,392; October 11, 1966; assigned to Yeda Research and Development Co., Ltd., Israel has described a very simple and convenient method for the preparation of enzymatically active water-insoluble substances consisting of enzyme molecules bound via functional groups of the enzyme molecules which are not necessary to the enzymatic activity, to cellulose. The linkage is accomplished by first preparing bromoacetylcellulose (BAC) and by binding molecules of enzymes through sulfhydryl groups or through α- or ϵ- amino groups of the enzyme molecules, or through the imidazole group of histidine. The compounds can be represented by the general formula

(1)
$$\text{cell} \left(\text{O} - \underset{\underset{\text{O}}{\|}}{\text{C}} - \text{CH}_2 - \text{X} - \text{Z} \right)_a$$

where cell designates cellulose, X designates a member of the group consisting of $-NH$, $-S-$, imidazole and Z designates a molecule of the enzyme whereof the $-S-$,

or

groups form part. The compounds of Formula 1 are prepared by reacting cellulose in any desired form (powder, sheets, etc.) with bromoacetyl bromide, resulting in a compound of the general formula:

$$cell + n(Br-CO-CH_2-Br) \longrightarrow cell-(O-CO-CH_2-Br)_n$$

which is subsequently reacted with a suitable enzyme, having functional groups adapted to result in a covalent bonding of the enzyme molecules without destruction of the enzymatic activity, resulting in compounds of the general Formula 1. The bonding reaction may be represented by:

$$cell\text{-}(-OCOCH_2Br)_n + mHX-Z \longrightarrow cell \begin{cases} -OCOCH_2Br)_{n-m} \\ -OCOCH_2X-Z)_m \end{cases} + mHBr$$

where X and Z have the same meaning as above.

Example 1: Preparation of Water-Insoluble Trypsin — 1 mg trypsin was stirred for 3 minutes with 1 gram of BAC (previously homogenized), in 50 ml of Veronal buffer (0.06 molar) of pH 8.6. The preparation (Tryp-BAC) was centrifuged off and washed 7 times with water. Almost all of the trypsin was bonded to the carrier.

In another experiment coupling of trypsin with BAC was carried out in 45% solution of acetone in Veronal buffer of pH 8.6. Total nitrogen analysis of the water-insoluble derivative (Tryp-BAC) showed that 100 mg of BAC bound 27 mg of trypsin.

Example 2: Preparation of Carrier-Bound Chymotrypsin — 500 mg of homogenized BAC were stirred for 4 minutes with 10 mg of chymotrypsin dissolved in 40 ml of Veronal buffer (0.06 molar) of pH 8.6 and 60 ml of water. The preparation (Chym-BAC) was centrifuged off and washed 7 times with water. Chym-BAC was found to be 45% active towards L-tyrosine ethyl ester. Enzymatic activity of Chym-BAC did not deteriorate during 3 months of storage coupling was followed by treatment with enough mercapto-ethanol to react with uncoupled bromo groups.

Example 3: Preparation of Carrier-Bound Ribonuclease — To 50 mg of BAC in 10 ml 0.1 molar sodium phosphate there was added 10 mg ribonuclease. The reaction mixture was stirred during 16 hours at a temperature of 4°C. After washing five times with 0.2 molar sodium acetate (pH 6.0) and twice with water, the remaining bromine groups were removed by reaction with mercaptoethanol. The enzymatic activity was tested towards cytidine 2',3'-cyclic phosphate and RNA, and compared with native ribonuclease. The activity towards the first substrate was equivalent to that of 750γ of the native enzyme, while towards the second substrate it was equivalent to 600γ.

The activity of the insoluble enzymatic preparation did not deteriorate during 3 months of storage, provided coupling was followed by treatment with enough mercaptoethanol to react with uncoupled bromo groups.

Triazinylcellulose Derivatives

The utility of enzymes as catalysts and other biologically active substances such as antigens, antibodies, etc. for example, in purifications, may be increased by attaching them to solid supports so that, for example, they may be removed from the reaction mixture or used in processes in which the reactants flow continuously over them. Also, the stability of an anzyme attached to a solid support is often greater than that of the free enzyme.

E.M. Crook and G. Kay; U.S. Patent 3,619,371; November 9, 1971; assigned to National Research Development Corporation have described a polymeric matrix having a biologically active substance chemically bound thereto, which comprises a polymer and a biologically active substance linked by groups of the formula:

$$
\begin{array}{c}
\text{N} \\
\text{C} \diagup \diagdown \text{C} \\
\text{N} \diagdown \qquad \diagup \text{N} \\
\text{C} \\
\text{Y}
\end{array}
$$

where Y represents a nucleophilic substituent that is an amino group, or an aliphatic or aromatic group.

The process for producing a polymeric matrix having a biologically active substance chemically bound thereto comprises reacting the substance with a polymer having attached thereto groups of formula:

$$
\begin{array}{c}
\text{N} \\
-\text{C} \diagup \diagdown \text{C}-\text{X} \\
\text{N} \diagdown \qquad \diagup \text{N} \\
\text{C} \\
\text{Y}
\end{array}
$$

where X represents a radical capable of reacting with the biologically active substance, e.g., a halogen atom, and Y represents a nucleophilic substituent that is an amino group, or an aliphatic or aromatic group.

The group Y is preferably one that carries a positive charge when in contact with solutions having a pH in the normal biological range between about pH 2 and 10, particularly between 5 and 9. Groups that are electrically neutral or that carry a negative charge can be used in some circumstances, but it has been found that the presence of such a positive charge frequently assists the reaction of a biologically active substance with the polymer.

Where Y is a nucleophilic aliphatic or aromatic substituent group, this is preferably one which contains nitrogen, oxygen or sulfur and particularly a nitrogen-containing group such as, for example, a substituted amino group.

As examples of preferred nucleophilic groups there may be mentioned amino, alkyl or aryl substituted amino, alkylamino, arylamino, oxyalkyl, oxyaryl, thioalkyl, and thioaryl groups.

Where X is a halogen atom this is preferably a chlorine or fluorine atom. The reaction between the polymer and the biologically active substance occurs by the replacement of the halogen atom represented by X by the residue of the biologically active substance. The reaction may be carried out by bringing the polymer and the biologically active substance into contact in an aqueous medium, for example, by adding the polymer to an aqueous solution of the biologically active substance. Since biologically active substances are unstable, the temperature should preferably be below, for example, room temperature and the pH as near to neutral as possible. However, it may be necessary to use a slightly alkaline pH, for example, between pH 7 and pH 8.6, as the reaction occurs more rapidly at alkaline pH. At neutral pH it may not be sufficiently rapid for practical purposes.

Polymers having groups of the given formula wherein Y represents a halogen atom can be prepared. These can be converted into polymers by the use of a suitable reagent to replace one of the halogen atoms with a specified nucleophilic group. Specific examples of suitable reagents include:

$$H_2N(CH_2)_2 \cdot N(C_2H_4OH)_2, \quad H_2N \cdot (CH_2)_2NH_2,$$

$$NH_3, \quad H_2NC_2H_5, \quad H_2H-\hspace{-0.3em}\bigcirc\hspace{-0.3em}-SO_3H,$$

$$HS \cdot C_2H_5, \quad HO-\hspace{-0.3em}\bigcirc\hspace{-0.3em}\cdot, \quad HS-\hspace{-0.3em}\bigcirc$$

This is a very convenient process for making polymers since the reagent also removes any traces of free cyanuric chloride which may be present in the polymer due to its method of preparation.

An alternative process for making polymers comprises reacting an s-triazinyl compound of formula:

$$
\begin{array}{c}
\text{N} \\
X-C \diagup \quad \diagdown C-X \\
\| \quad \quad \| \\
\text{N} \quad \quad \text{N} \\
\diagdown C \diagup \\
\| \\
Y
\end{array}
$$

where X and Y have the meaning given above, with a polymer having nucleophilic groups, and stopping the reaction before substantial cross-linking occurred.

Suitable polymers are often hydrophilic polymers, for example, those having free hydroxyl groups, and there can be used naturally occurring polymers such as celluloses, cross-linked dextrans, starch, dextran, and proteins such as wool, and synthetic polymers such as polyvinyl alcohol. The reaction of the polymer with the s-triazinyl compound can be carried out in aqueous solution provided that Y contains a solubilizing substituent, and the reaction can then be conveniently stopped by reducing the pH, for example, by washing with water. Suitable solubilizing substituents include, for example, carboxy groups, sulfonic acid groups, or $-N(alkyl)_3$ groups. If the s-triazinyl compound of the formula given above is not water-soluble, a mixed organic/aqueous solvent should be used. The compound of the above formula is not water-soluble when Y represents, for example, methoxy, ethylamino, butylthio or phenoxy.

Where Y contains an alkyl residue this is preferably a short chain residue since the cyanuric derivative of such a residue is more soluble than that of a long-chain residue. If long-chain residues such as C_6 are used it may be necessary to use a mixed organic/aqueous solvent even though the residue is substituted with a solubilizing group.

The reaction of the polymer with the s-triazinyl compound, the reaction of the s-triazinyl substituted polymer with the biologically active substance and the re-action of the polymer-bound biologically active substance in use may be controlled by selecting Y so that the reactants in the various reactions are appropriately charged.

Example 1: (a) Preparation of Carboxymethylaminochloro-s-Triazinylcellulose — 7.2 grams of sodium bicarbonate was dissolved in a mixture of 60 ml normal glycine solution, 50 ml distilled water and 50 ml acetone. This solution was added to a solution of 3.4 grams cyanuric chloride in 100 ml acetone. The mixture was vigorously stirred and much carbon dioxide was evolved.

After 1 minute the pH was reduced to 2.5 by addition of dilute hydrochloric acid, the carbon dioxide evolved being blown off by a stream of nitrogen. The acetone was then removed by evaporation under reduced pressure at a temperature not greater than 40°C. The crystals which precipitated were filtered off, washed with a little distilled water and dried in a vacuum desiccator over silica gel for 2 hours.

The product was recrystallized by dissolving in a 50% (v/v) water-acetone mixture, and removing the acetone at low temperature. The 2,4-dichloro-6-carboxy-methylamino-s-triazine was added to 5 ml distilled water, and the pH of the mixture raised to 6 by addition of 4N sodium hydroxide solution. At this pH the s-triazine compound dissolved completely. The total volume of solution was then made up to 10 ml by addition of distilled water and 2 grams of cellulose powder was added. The pH was raised to 11 and maintained at that (±0.2 of pH unit) by the dropwise addition of 4N sodium hydroxide. A sample was taken 10 minutes after the pH had been adjusted to 11. The reaction was carried out in a water bath at 25°C.

However, because the reaction was exothermic the temperature of the reaction mixture rose to a maximum of 30°C, at 3 minutes after which it fell again to 26°C, at 10 minutes. The sample was pipetted into 20 ml dilute acetic acid (20% v/v), filtered, washed with deionized water until the filtrate was chloride-free, as measured with silver nitrate, and then stirred overnight in 10 ml deionized water, and dried in a vacuum desiccator over silica gel.

(b) Preparation of Carboxymethoxychloro-s-Triazinylcellulose — 5 grams glycolic

acid and 5 grams sodium bicarbonate were stirred into 10 ml distilled water until they both completely dissolved. This solution was added to a solution of 5 grams cyanuric chloride in 20 ml acetone. The mixture was stirred and the temperature maintained at less than 30°C by addition of small amounts of crushed ice. After 20 minutes this mixture, which now had become a clear solution, was put on to the rotary evaporator, and the excess acetone was removed under reduced pressure at less than 30°C. The 2,4-dichloro-6-carboxymethoxy-s-triazine was precipitated out as a thick oil.

The aqueous phase was poured off the oil, and the oil was washed with about 5 ml of distilled water. The 2,4-dichloro-6-carboxymethoxy-s-triazine was redissolved by adding 10 ml distilled water containing excess (2.4 grams) sodium bicarbonate. The final volume, plus washings, was 15 ml. 3 grams of cellulose powder was then added, and the mixture stirred at 25°C in a water bath. The pH of the solution was raised to 9 and maintained at that by the addition of 4N sodium hydroxide solution.

The buffering effect of the carbonate ions allowed this to be done without difficulty. Samples were taken 6 and 15 minutes after the start of the reaction. The samples were pipetted in 20 ml dilute acetic acid (20% v/v), filtered, washed with deionized water until the filtrate was chloride-free, as measured with silver nitrate, and then stirred overnight in 10 ml deionized water. Finally the cellulose was again filtered off, washed with a further 10 ml deionized water, and dried in a vacuum desiccator over silica gel.

Reaction of Triazinylcellulose Derivatives with Chymotrypsin — The amount of enzyme attached to a cellulose sample was estimated as the difference between the amount of enzyme protein added initially and the amount removed by subsequent washing as estimated by the method described by O.W. Lowry, N.J. Rosebrough, A.L. Farr and R.J. Randell, *J. Biol. Chem.* 193, 265 (1951).

The washing was carried out by placing the solution containing the cellulose and chymotrypsin on a Buchner funnel and removing the excess liquid by gentle suction. The residue was washed with a solution which was 1N in sodium chloride and 2M in urea. The filtrate was collected and the concentration of enzyme was estimated by the Lowry et al method, a 2M urea 1N sodium chloride solution being used as a blank. About 95% of the adsorbed enzyme is removed by this method. The enzyme activity was measured semiquantitatively as follows:

The reaction was carried out in 5 ml of a solution 0.50N in sodium chloride and 0.00112M in sodium phosphate. The pH of the solution was 8.00 ± 0.02. The enzyme-cellulose, or 0.1 ml of the free enzyme solution, was added to this solution, and after readjustment of the pH, 0.25 ml of N-acetyl-L-tyrosinic ethyl ester monohydrate (ATEE) solution was added.

The ATEE solution contained 0.0574 grams ATEE dissolved in 4 ml spectrosol dioxane and made up to 25 ml total volume with deionized water. Thus the final assay solution contained 2.02×10^{-6} mols of ATEE, and is therefore

$$6.02 \times 10^{-3} \text{ M}$$

It contained 0.75% (v/v) organic solvent. The reaction was followed titrimetrically in the usual way, using 0.01N sodium hydroxide as titrant. In the following results, percent activity means 100 x activity of a given weight of attached enzyme/activity of the same weight of unattached enzyme.

Reactions of (a) Carboxymethylaminochloro-s-Triazinylcellulose and (b) Carboxymethoxychloro-s-Triazinylcellulose with Chymotrypsin — The experiments were carried out as follows.

A small weight of the cellulose sample was stirred for 5 minutes in 0.01N hydrochloric acid, filtered and washed until chloride-free. It was then put in 1 ml enzyme solution. 1 ml phosphate buffer solution was then added. The mixture was then stirred overnight either in the cold room, when the sample temperature was $2° \pm 0.5°C$, or in a water bath at 25°C. The excess enzyme was then washed off, and the degree of substitution and amount of activity remaining estimated in the usual way.

s-Triazinylcellulose Used	(b)*	(a)	(b)**
Weight of sample, g	0.2907	0.2746	0.2015
Weight of enzyme, g	0.0082	0.0082	0.0082
Strength of phosphate buffer	N/5	1N	1N
Temp, °C	2	25	25
pH	8.6	8.6	8.6
Percent substitution	0.765	0.46	1.15
Percent activity	46	19	17

*6-minute sample.
**15-minute sample.

Example 2: Twelve discs of No. 1 paper (2.5 cm diameter) were soaked in 1N NaOH for 5 minutes. Excess NaOH was dampened off and the papers stirred in dioxane (20 to 30 ml) for 5 minutes and then removed. The papers were then added to 20 ml of a solution of cyanuric chloride (0.25 g/ml) at room temperature, followed immediately by 25 ml water and then after about 5 seconds by 20 to 30 ml, glacial acetic acid. The papers were then removed and resuspended in 20 ml dioxane. After 2 minutes 20 ml water and a little glacial acetic were added. After a further 5 minutes the papers were removed and washed thoroughly on a Buchner funnel with water and then with acetone. Finally they were dried in a vacuum desiccator and stored at 2°C.

A solution of N-(3-aminopropyl)diethanolamine (3 g/100 ml) was brought to pH 7 with HCl. 10 ml water was added to 10 ml of this amine solution and the papers suspended in it for 7½ minutes at room temperature. 10 ml of 1N HCl was then added to stop the reaction and the papers were then washed with 5M NaCl followed by water and finally acetone. They were then replaced in the desiccator.

Attachment of β-Galactosidase — The enzyme (0.1 ml) plus EDTA (0.001M) 0.025 ml plus $MgSO_4$ (0.1M) 0.025 ml was added to the dried paper (one disc) and left for one-half hour. After washing it was found that three discs gave a good optical density change 0.5 at 420 mμ when reacted on ortho-nitrophenyl galactoside at 4 ml/min. The excess active chlorine and any second chlorine were removed by leaving the discs in the N-(3-aminopropyl)diethanolamine solution with small amounts of EDTA and $MgSO_4$ solutions.

Attachment of Pyruvate Kinase — Pyruvate kinase (0.25 ml of 5.5 mg/ml preparation in $(NH_4)_2SO_4$ was dialyzed overnight at 4°C, against 2 liters of buffer containing 3 mmols potassium phosphate, pH 7.4, 0.03 mmol EDTA and 0.03 millimol EDTA and 0.375 mmol $MgSO_4$. The dialyzate (0.4 ml) was diluted with 1 ml of a solution of potassium phosphate (100 mmols, pH 7.4) and 0.1 milliliter of this mixture was then pipetted on to each of six discs. The discs were covered and left for 50 minutes at about 25°C. They were then suspended in 15 to 20 ml of the amine solution containing:

> 30 mmols N-(3-aminopropyl)diethanolamine
> 2.1 mmols EDTA
> 11 mmols $MgSO_4$
> 42 mmols potassium phosphate

The pH of the solution was 7.8. After 1 day at room temperature, the discs were stored in a refrigerator. One such disc gave a fair degree of reaction when substrate was passed through at 4 ml/minute.

Dextran and Agarose Derivatives

R.E.A.V. Axen, J.O. Porath and E.S. Ernbach; U.S. Patent 3,645,852; February 29, 1972; assigned to Pharmacia AB, Sweden have developed a method of binding, by covalent bonds, water-soluble proteins and water-soluble peptides containing one or more groups of the formula −YH, in which −YH represents a primary or secondary amino group, to water-insoluble polymers containing one or more groups of the formula −XH, in which −XH represents a hydroxyl group or a primary or secondary amino group. The group −XH as well as the group −YH contains a reactive hydrogen atom H.

The water-insoluble polymer containing one or more groups of the formula −XH is reacted with a cyanogen halide and with the water-soluble protein or water-soluble peptide. The reaction is carried out in an aqueous weakly alkaline medium. It is thus carried out in the presence of water, which is of special significance, and the temperature is variable, for example, in the range of from 0° to 50°C. The reaction is based on the formation of bridges with covalent bonds between the water-insoluble polymer and the water-soluble protein or peptide. Investigations carried out have revealed that the bridges have the formula

$$A-X-\underset{\underset{Z}{\|}}{C}-Y-B$$

wherein A is the residue of the water-insoluble polymer and B is the residue of the protein or peptide and Z stands for imino(=NH) or oxygen (=O). That Z can also stand for oxygen is due to the fact that the imino group can in certain cases convert into oxygen by hydrolysis.

The reaction can be carried out in two stages. The first of these stages is commonly effected in such a manner that the water-insoluble polymer containing one or more groups having the formula −XH is contacted with a cyanogen halide, the latter being present in the form of the pure substance or a solution thereof. As a cyanogen halide, there may usually be used the iodo-, chloro-, or bromo-compound or optionally a mixture thereof. The reaction is carried out under alkaline conditions, which are provided by the addition of a suitable alkaline reacting substance to the aqueous reaction medium, such as sodium hydroxide. Suitable pH values are primarily such values in the range of from 8 to 13. If −XH is a primary or secondary amino group, it is possible to work in a more weakly alkaline medium than when −XH is a hydroxy group.

The reaction can be carried out at different temperatures, for example, in the range of from 0° to 50°C. When working in aqueous solution, certain losses of cyanogen halide arises due to part of the latter being consumed by hydrolysis. The reaction results in the formation of reactive derivatives of the water-insoluble polymer which can then immediately or optionally after isolation, be further reacted with the water-soluble protein or water-soluble peptide. The reactive derivative as formed can also be stored under suitable conditions and later be reacted with the biopolymer. Prior to being bound to the biopolymer, a reactive derivative based on a water-insoluble polymer can be purified, for example by washing.

The second stage of the process is preferably carried out under weakly alkaline conditions at a temperature which can, for example, be in the range of from 0° to 50°C, such as at room temperature. It is an advantage that this stage can be carried out in the presence of water.

The water-insoluble polymer can be a water-insoluble polysaccharide or a water-insoluble derivative of a polysaccharide which contains hydroxyl groups and/or primary and/or secondary amino groups. An example of a water-insoluble polysaccharide is cellulose. Examples of insoluble polysaccharide derivatives are cross-linked polysaccharides, for example, cross-linked dextran, starch, dextrin, and hydroxyl group-containing derivatives of such polysaccharides such as hydroxyethyl group-containing derivatives or amino group-containing derivatives such as p-aminophenoxyhydroxypropyl group-substituted cross-linked dextran and starch. Another example is agarose. A still further example is water-insoluble cross-linked polyvinyl alcohol and synthetic polymers containing amino groups.

The polymer is insoluble in water, but it can be swellable in water. In this connection, it can consist of a cross-linked three-dimensional hydroxyl group-containing network which is insoluble but swellable in water. As examples of such polymers may be mentioned copolymers of dextran with a bifunctional substance, such as epichlorohydrin, such copolymers being characterized by presenting a varying ability of absorbing water. Copolymers of other polysaccharides or other hydroxyl compounds with bifunctional substances may come in question, such as copolymers of saccharose and of sugar alcohols, such as sorbitol, mannitol, and polyvinyl alcohol. As copolymers may also be mentioned those which are substituted by primary or secondary amino groups such as p-aminophenoxyhydroxypropyl groups.

As water-soluble proteins and peptides, containing one or more groups of the formula $-YH$, which are to be subjected to the binding reaction with the reactive derivative may be mentioned enzymes, antibodies, protein and/or peptide hormones or antigenic proteins.

Example 1: This illustrates the binding of chymotrypsin to a copolymer of dextran with epichlorohydrin. (A) Activation of the copolymer by reacting with cyanogen bromide – 200 mg copolymer of dextran with epichlorohydrin (Sephadex G-200) a water-insoluble gel product which is swellable in water were combined with 8 ml of an aqueous solution containing 25 mg of cyanogen bromide per ml of water. The reaction was allowed to take place at 23°C while stirring and the pH of the mixture was maintained at 11.5 using an automatic titrator adding 2M sodium hydroxide solution. The reaction time was 6 minutes. The product was washed rapidly with suction on a glass filter with cold water, and used for the next stage of the reaction.

(B) Binding with chymotrypsin – The activated copolymer obtained according to (A) was contacted with 200 mg of chymotrypsin in 2 ml of 0.1M sodium hydrocarbonate solution, whereupon the mixture was allowed to react at 23°C for 30 hours while being stirred. Subsequent to being subjected to suction on a glass filter the obtained product was washed with 0.1M sodium hydrocarbonate solution, 10^{-3}M hydrochloric acid, water, 0.5M sodium chloride solution, water and a water-acetone mixture with increasing concentrations of acetone, and finally with pure acetone. The shrunken product was then dried. The result obtained upon analysis of amino acids was a content of 200 mg of protein per gram of dried final product. The product obtained has enzymatic activity and is used, for example, for analytical procedures.

Example 2: This illustrates the binding of glucose oxidase to agarose. (A) Activation of agarose by reaction with cyanogen bromide – Water-swollen, ball-shaped agarose particles (Sepharose 2B) were subjected to suction on a glass filter. 3.9 grams (corresponding to 100 mg of shrunken and dried agarose) were added with 4 ml of a cyanogen bromide solution containing 25 mg cyanogen bromide per ml of water, whereafter the activation process was effected analogously with Example 1(B) at pH 11 by adding 2M sodium hydroxide solution for 6 minutes using an automatic titrator at 23°C. The activated product was washed on a glass filter with 1 liter of ice water and was finally washed rapidly with 0.1M phosphate buffer having a pH of 7.4.

(B) Binding with glucose oxidase – The activated polymer was reacted with 24 mg of glucose oxidase which was dissolved in 1 ml of 0.1M phosphate buffer at pH 7.4. The binding process took place for 10 hours with slow rotation of the reaction vessel. The gel conjugate was placed in a column and washed with 1 liter of 0.1M acetate buffer of pH 4.6, this amount being pumped by means of a pump through the column for 10 hours.

A portion of the gel was removed, shrunken with acetone, dried and analyzed with respect to protein. The amount of bonded glucose oxidase was calculated therefrom to reach 30 mg per gram of dried polymer conjugate.

The bound enzyme was found to have considerable activity, and the catalytic column could be used for determining glucose in samples. The column was also found to possess good chromatographic properties. The water-insoluble product is a valuable reagent for analytical procedures.

Polyethylene Glycol Derivative of Cellulose; Activated Amino Acid Polymers

E. Katchalski and E. Bar-Eli; U.S. Patent 3,167,485; January 26, 1965 have found that enzymatically active, repeatedly usable modified enzymes can be prepared by bonding covalently via groups of the enzyme which are nonessential to their enzymatic activity either directly or through polymeric chains to water-insoluble carriers and by covalent bonds to chain-formed polymeric links, the latter being covalently bonded to preferably polymeric water-insoluble carriers.

Water-insoluble modified enzymes can be used repeatedly for effective enzymatic processes. Such processes can be carried out by introducing such water-insoluble modified enzymes into solutions of the substrate to be hydrolyzed, stirring the solution for a predetermined duration and filtering off.

Enzymatic processes can be effected by passing a solution of the substrate to be converted through columns of or columns containing such water-insoluble substrate in admixture with suitable inert volume extenders, the rate and degree of conversion being a function of the parameters of the column and of the rate of flow through the column.

As described in greater detail in Example 1, it has been found that a water-insoluble carrier, namely a copolymer of L-leucine and p-aminophenyl-DL-alanine can be prepared by reacting the corresponding N-carboxyanhydrides and that same can be bonded covalently to papain so as to result in water-insoluble modified enzyme which is enzymatically active. As shown in Example 2, the same polymeric carrier can be bonded covalently to polytyrosyltrypsin, prepared by reacting N-carboxy-L-tyrosine anhydride and trypsin, by preparing the diazonium salt of the water-insoluble carrier and by coupling same with the polytyrosyltrypsin.

In some cases, an enzyme such as papain, can be bonded directly to a water-insoluble carrier, without substantial loss of enzymatic activity. In other cases, as for example trypsin, chymotrypsin and urease, an intermediate link must be resorted to, as will be exemplified in the following.

Example 1: Water-Insoluble Papain — (a) Preparation of the water-insoluble carrier (copolymer I): The water-insoluble carrier was prepared by the copolymerization of N-carboxy-L-leucine anhydride with p-(N-carbobenzoxyamino)-N-carboxy-DL-phenylalanine anhydride and the removal of the carbobenzoxy groups from the copolymer thus obtained with anhydrous hydrogen bromide. It consists of a copolymer of L-leucine and p-aminophenyl-DL-alanine.

3.5 grams of p-(N-carbobenzoxyamino)-N-carboxy-DL-phenylalanine anhydride, prepared according to M. Sela and E. Katchalski, *J. Am. Chem. Soc.* 76,129 (1954), and 1.5 grams of N-carboxy-L-leucine anhydride, prepared according to A.C. Farthing, *J. Chem. Soc.* 3213 (1950), were dissolved in 100 ml dioxane and 0.1 ml dioxane and 0.1 ml of triethylamine was added. The reaction mixture was kept at room temperature with magnetic stirring for 72 hours. The solution was refluxed during 1 hour in an oil-bath and cooled down to 25°C. Precipitation of the copolymer was effected by addition of distilled water, yielding 3.5 grams of a yellowish substance. This substance was dried over phosphorus pentoxide and the carbobenzoxy groups were removed by addition of 30 ml of a 33% HBr/acetic acid solution at room temperature, according to D. Ben-Ishai and Berger, *J. Org. Chem.*, 17, 1564 (1952). Dissolution took place within 2 hours and precipitation was effected by the addition of anhydrous ether. The excess of HBr was removed by repeated washing with ether and centrifugation. The thus obtained substance was dried over sodium hydroxide. Yield: 2.5 grams.

The copolymer of p-(N-carbobenzoxyamino)-N-carboxy-DL-phenylalanine and L-leucine was analyzed for amino nitrogen content. From the value obtained a number average molecular weight of 3,400 was calculated. A chromatographic analysis of an acid hydrolyzate (6N hydrochloric acid) of the copolymer I yielded p-aminophenyl-DL-alanine in a molar ratio of 1 to 2.5.

(b) Coupling of papain with copolymer I: The copolymer (250 mg) was dissolved in 50% aqueous acetic acid (3 ml) and the solution was mixed with 2N HCl (3 ml). Diazotization was effected by dropwise addition of 0.5 ml sodium nitrate solution to the ice-cooled polymer solution. The polydiazonium salt was precipitated after two hours at 4°C by adding 2N NaOH to pH 8.0. It was washed twice with 12 ml portions of 10% sodium acetate followed by 0.1M sodium phosphate buffer, pH 7.6 (30 ml). Coupling with papain was carried out by adding crystalline papain (80 mg) suspended in 0.03M cysteine (2.0 ml) to a suspension of the diazotized copolymer in an aqueous solution (40 ml) at pH 7.6, 0.075M in phosphate, 0.005M in cysteine and 0.002M in ethylenediamine tetraacetate.

Oxygen was removed by flushing the mixture with nitrogen and the coupling was allowed to proceed in a tightly stoppered vessel for 20 hours at 4°C, with magnetic stirring. No free diazonium groups could be detected by means of α-naphthol after this period. The water-insoluble coupling product was centrifuged down and washed several times with a solution at pH 6.0, 0.005N in cysteine and 0.002N in ethylenediamine tetraacetate. No enzymatic activity was detected in the initial supernatant or in any of the subsequent washings.

The final water-insoluble papain (250 mg) was stored at 4°C under an aqueous solution similar in composition to the wash solution. To facilitate pipetting of aliquots of constant enzymatic activity, the water-insoluble enzyme was homogenized in a Potter homogenizer with a Teflon piston. No loss in activity occurred during this procedure.

Example 2: Water-Insoluble Trypsin — (a) Preparation of the water-insoluble carrier (copolymer I): A copolymer of L-leucine and p-amino-DL-phenylalanine was prepared as described in Example 1(a) above.

(b) Preparation of polytyrosyltrypsin (PTT): 155 mg of trypsin, containing 50% by weight of magnesium sulfate, were dissolved in 12 ml phosphate buffer of pH 7.2 and ionic strength 0.25. The solution was cooled to 0°C and 130 mg of N-carboxy-L-tyrosine anhydride, prepared according A. Berger et al, *Bull. Res. Council Israel,* 7A, 98 (1958), dissolved in 3 ml dioxane were added dropwise resulting in a slightly turbid solution. Magnetic stirring was effected at a temperature of 2°C during 20 hours. The solution obtained was dialyzed during 5 days against 0.0025N hydrochloric acid at a temperature of 2°C, with 5 changes of the dialysis fluid. The clear solution obtained was lyophilized, yielding 75 milligrams polytyrosyltrypsin. The absence of free tyrosine in the enzymatic preparation obtained was proved chromatographically.

The quantity of tyrosine which reacted with trypsin to form the polytyrosyltrypsin (PTT) preparation obtained was derived from the UV adsorption in 0.1N sodium hydroxide at $\lambda = 293.5$ mμ, using the corresponding quantity of trypsin as a blank. The PTT was found to contain 8.5% by weight of tyrosine in the form of tyrosine peptide side-chains.

(c) Coupling of polytyrosyltrypsin (PTT) with copolymer I: 200 mg of copolymer I were dissolved in a mixture of 4 ml of water and 0.8 ml 2N hydrochloric acid and the mixture was cooled to -2°C. 0.5 ml of 0.5M sodium nitrite was added within 1 hour with occasional stirring. The diazonium salt of the copolymer I was precipitated by adding 5 ml of 10% aqueous sodium acetate and raising the pH to 8 by adding 1N NaOH. The precipitate was centrifuged and washed with 10% sodium acetate and with phosphate buffer of pH 7.3 ionic strength 0.25, yield of diazonium salt 160 mg.

To the polymeric diazonium salt obtained were added 3 ml phosphate buffer, pH 7.3, ionic strength 0.25 and a solution of 29 mg polytyrosyltrypsin in 2 ml of 0.0025N hydrochloric acid. The resulting mixture was kept at 2°C for one and a half hours, the pH was adjusted to 6.8 with 0.1N HCl and the mixture left for an additional 18 hours with magnetic stirring at 2°C.

The final water-insoluble complex obtained was washed with cold phosphate buffer of pH 7.3, ionic strength 0.25, 10% sodium acetate and 0.0025N hydrochloric acid. A yield of 130 mg of a water-insoluble polytyrosyltrypsin copolymer (designated as I-PTT) was thus obtained.

The protein content (20% by weight) of the I-PTT synthesized was calculated from the amount of valine obtained on total acid hydrolysis. Quantitative chromatographic analysis of the acid hydrolyzate yielded 1.15% of valine for I-PTT. Trypsin yielded under similar conditions 5.7% of valine.

(d) Estimation of active enzyme in I-PTT: The enzymatic activity of I-PTT towards benzoyl-L-arginine methyl ester (BAME) was determined titrimetically using a pH-stat titrator model TTT-1, with an automatic recorder. The initial rate of hydrolysis of the 9 mg of the ester BAME in 2.5 ml phosphate buffer 0.01M at pH 7.8 by 1.7 mg of I-PTT, corresponded to that of 0.083 mg of crystalline trypsin. It may thus be concluded that the tryptic activity of 1,000 milligrams I-PTT towards BAME corresponds to that of 50 mg crystalline trypsin.

(e) Preparation of a trypsin column: The water-insoluble I-PTT (6 mg) was mixed with 500 mg Geon (inert vinylic resin 426 Goodrich) and the mixture introduced into a glass tube of 6 mm internal diameter and 100 mm height.

(f) Illustration of the possible use of the trypsin column: The rate of hydrolysis of L-arginine methyl ester by the column when the ester was passed at different rates was determined as follows. A solution containing 3.5 mg of L-arginine methyl ester per ml 0.1M phthalate buffer pH 6.2, ionic strength 0.19, was prepared. The solution was passed through the column at different rates by changing hydrostatic pressure and the amount of intact ester determined by Hestrin's procedure, S. Hestrin, *J. Biol. Chem,* 180, 249 (1949). The extent of hydrolysis is inversely proportional to the rate of flow through the column, as evident from the table below.

Rate of Flow	Hydrolysis of Ester, %
1 cc/ 2 min	20
1 cc/ 5 min	50
1 cc/ 8 min	80
1 cc/10 min	100

Experiments have shown that such columns can be used repeatedly without a perceptible decrease of the enzymatic activity.

Enzymatic hydrolysis of poly-L-lysine could also be effected by the column. Thus when a solution of poly-L-lysine hydrobromide (10 mg per ml 0.05M phosphate buffer, pH 7.6) was passed through the trypsin column at the rates given in the table below, the following products could be determined chromatographically in the effluent: lysine, dilysine, trilysine, and tetralysine, according to the procedure described by S.G. Waley and J. Watson, *Biochem. J.,* 55, 328 (1953). The amounts of these digestion products as a function of the rate of flow are presented in the table below.

Rate of Flow	Lysine, mg/cc	Dilysine, mg/cc	Trilysine, mg/cc	Tetralysine, mg/cc
1 cc/ 26 min	0.09	1.94	3.22	2.0
1 cc/ 60 min	0.164	2.21	3.50	1.53
1 cc/120 min	0.38	3.15	3.15	0.425

Protamine sulfate could also be hydrolyzed by passing through the trypsin column. The digestion products determined chromatographically were found to be identical with those formed on incubation with native trypsin.

Example 3: Water-Insoluble Urease — It is possible to prepare modified, water-insoluble urease by bonding this enzyme covalently via its sulfhydryl groups nonessential to its enzymatic activity by means of modified polyethylene glycol to cellulose.

Polyethylene glycol of a molecular weight of about 200 is subjected to esterification of one of the terminal hydroxyl groups by means of iodoacetic acid, resulting in iodoacetic acid polyethylene glycol monoester. The second terminal hydroxyl group is converted into a chloroformate substituent by reacting it with phosgene, and this reactive chloroformate group is reacted with hydroxyl group of cellulose resulting in its bonding thereto with the splitting off of HCl. The terminal iodoacetic radical is reacted with the sulfhydryl grouping of the enzyme, and subsequently the polyethylene ester chain is bonded thereto with a simultaneous evolution of hydrogen iodide. An amount of 4 grams polyethylene glycol, MW 200, was subjected to esterification by heating with 3 grams iodoacetic acid (ICH_2COOH) for 2 hours at 200°C. Yield of liquid iodoacetic acid polyethylene glycol monoester was 6 grams. Liquid phosgene (1.5 ml) was added dropwise to the ice cooled monoester and the reaction mixture stirred for one hour. Excess of phosgene was removed with nitrogen. One gram of the chloroformate of the monoester was then added to 25 grams of cellulose powder in a 10% solution of sodium bicarbonate, and the mixture stirred for 3 hours at 0°C.

The water-insoluble product was washed with water and dried in vacuo. The final product contained 0.5% organic iodine. The cellulose derivative obtained (200 mg) was suspended in 6 ml of 0.1M phosphate buffer, pH 7.6, and a solution of 30 mg of urease in 4 ml of the same buffer was added. Stirring was effected at 2°C for 20 hours.

The insoluble modified urease formed was separated by centrifugation and washed three times with 0.1M phosphate buffer, pH 7.6. The water-insoluble preparation showed urease activity equal to about 3% by weight respective an equal weight of crystalline urease.

Cellulose Carbonate

In a process of *S.A. Barker, J.F. Kennedy and C.J. Gray; U.S. Patent 3,810,821; May 14, 1974; assigned to Ranks Hovis McDougall Limited, England*, cellulose carbonate is prepared by reacting cellulose or a 6-substituted cellulose, e.g., methylcellulose, hydroxyethylcellulose, carboxymethylcellulose and diethylaminoethylcellulose with an alkyl or aryl chloroformate. Water-insoluble enzyme preparations are produced by reacting the enzyme, e.g., mouse kidney extract, β-glucosidase, trypsin, α-amylase, glucoamylase or mouse liver extract, dissolved or suspended in aqueous solution with the cellulose carbonate.

The reaction of excess of ethyl chloroformate with pyranoid compounds containing vicinal diequatorial hydroxy groups is known to give trans-five membered cyclic derivatives in the presence of triethylamine. Thus, methyl 4,6-O-benzylidene-α-D-glucopyranoside gave the 2,3-carbonate and methyl 2,6-di-O-methanesulfonyl-α-D-glucopyranoside gave the 3,4-cyclic ester [W.M. Doane, B.S. Shasha, E.I. South, C.R. Russell and C.E. Rist, *Carbohyd. Res.* 4,445 (1967)]. Depending upon reaction conditions methyl 4,6-O-benzylidine-α-D-glucopyranoside also gave the 2,3-di-O-ethyloxycarbonyl or the 2- and 3-monoesters. These workers also stated that the 2,3-trans fused cyclic carbonates showed characteristic absorptions in the region of 1,825 cm^{-1} and 1,840 and 1,810 cm^{-1} of the infrared spectrum.

The O-ethyloxycarbonyl group absorbed at 1,750 cm^{-1}. Carbonate derivatives have been made of polysaccharides such as dextran and dextrin, that are soluble in dimethyl sulfoxide [W.M. Doane, E.I. South, B.S. Shasha, C.R. Russell and C.E. Rist, *Carbohyd. Res.* 5, 366 (1967), 8, 266 (1968)]. Cellulose carbonate is the first insoluble polymer converted to its carbonate derivative. 2,3-trans cyclic carbonates undergo ring opening with methanol, toluenethiol or piperidine

to give mixtures of the 2- or 3-substituted derivatives in the presence of triethylamine [E.I. South, W.M. Doane, B.S. Shasha, C.R. Russell and C.E. Rist,
Tetrahedron Letters 4481, (1967)].

The process provides a cellulose carbonate more particularly a cellulose 2,3-car-
bonate and further provides derivatives of cellulose including trans cyclic carbonate groupings bridging C_2 and C_3 of some or all of its 1,4-β-D-glucopyranose
units. The cellulose may also be substituted at C_6 in some or all of its 1,4-β-D-
glucopyranose units. Thus the cellulose carbonate may contain not only cyclic
carbonate structures derived by reaction of ethyl chloroformate with the trans
hydroxyl groups on carbon two and carbon three of some or all of the D-glucopyranose residues in the cellulose but also O-ethyloxycarbonyl groups derived
by the reaction of ethyl chloroformate with certain of the hydroxyl groups on
carbon two, three or six of the D-glucopyranose residues.

The chloroformate is preferably ethyl chloroformate. Methyl chloroformate,
phenyl chloroformate and benzyl chloroformate may also be employed. The
reaction may be conveniently carried out at $0°C$ for approximately 10 minutes
in dimethyl sulfoxide/dioxane. Dimethylformamide/dioxane or other solvent mixture may also be employed.

A base for example a tertiary amine is preferably present. All components of
the reaction are preferably dry. The ethyl chloroformate and tertiary amine,
e.g., triethylamine, may be dissolved in dimethyl sulfoxide or dimethylformamide or other solvent or solvent mixture devoid of $-OH$, $-SH$, $-NH$ and $-NH_2$
groups.

The 6-substituted cellulose may be methylcellulose, hydroxyethylcellulose, carboxymethylcellulose or diethylaminoethylcellulose. Thus the cellulose employed
as starting material may be already substituted at C_6 in some or all of its 1,4-β-
D-glucopyranose residues.

The cellulose carbonate products may also be employed as intermediates in the
preparation of 6-substituted cellulose derivatives with modified solubility properties when derivatives may be more resistant to microbial breakdown.

Example 1: Cellulose (1 g) was suspended in liquid dimethyl sulfoxide (10 ml),
p-dioxane (1.5 ml) and triethylamine (8 ml) and stirred at $0°C$ for 5 minutes.
Ethyl chloroformate (16 ml) was added dropwise, and stirring continued for 10
minutes.

The mixture was then neutralized to pH 7.0 with 6N HCl. The mixture was
suspended in 90% ethanol (400 ml) at $20°C$ using a Waring blender, filtered,
and the cellulose carbonate washed with 90% ethanol (200 ml), ethanol (200 ml),
diethyl ether (200 ml) and dried in vacuo over P_2O_5 at $20°C$.

Example 2: Coupling of Cellulose Carbonate with Mouse Kidney Extract —
Washed mouse kidneys (4 grams) were macerated in a top-drive macerator with
0.1M citrate buffer, pH 4.5 (50 ml) for 5 minutes. The resulting mixture was
then centrifuged at 8,700 g for 20 minutes and the supernatant taken. Cellulose carbonate (Example 1, 200 mg) was suspended in water (10 ml) at $4°C$,

and mouse kidney extract (5 ml) in 0.05 citrate buffer, pH 4.4, added followed by triethylamine (0.2 ml). After stirring for several hours both this and a control in which cellulose replaced cellulose carbonate were centrifuged. The solids were washed with 0.05M citrate buffer, pH 4.4, (5 ml) and then M sodium chloride in the same buffer in five alternate sequences followed by three washings with buffer.

The final centrifuged products (MKE 1 and control 1) were suspended in 10 ml 0.05M citrate buffer, pH 4.4. The enzyme activity was assayed by incubating 1 ml MKE suspension or control 1 with 1 ml p-nitrophenyl-2-acetamido-2-deoxy-β-D-glucopyranoside (10 mg in 5 ml citrate buffer) at 40°C. Reaction was terminated by addition of an equal volume of 0.2M sodium carbonate, and the OD of the centrifuged supernatant read at 420 mμ (MKE 1, 0.206; cellulose control 1, 0.086; substrate control, 0.053). Thus, the enzyme β-N-acetyglucosaminidase is shown to have been bound to the cellulose.

Example 3: Coupling of Cellulose Carbonate with Almond β-Glucosidase – Cellulose carbonate, Example 1 (200 mg) was suspended in water (10 ml), and β-glucosidase (10 mg) in water (5 ml) added followed by triethylamine (0.2 ml). After stirring for 3 hours at 4°C, the solid was filtered off, washed with a solution containing both M sodium chloride and M sucrose and then a solution of 0.005M acetate buffer, pH 4.95, five times alternately, followed by the acetate buffer twice before being suspended in 0.005M acetate buffer, pH 4.95 (10 ml).

The product (BG 1), and cellulose control 2 carried through the same procedure were separately assayed by incubating 1 ml suspension with 1 ml O-nitrophenyl-β-D-glucopyranoside (0.1076 gram in 10 ml acetate buffer) at 37°C. The reaction was terminated after 3.5 hours and the OD of each centrifuged supernatant read at 420 mμ (BG 1, 2.235; cellulose control 2, 0.417; substrate control, 0.130).

In a repeat experiment but effecting the coupling with β-glucosidase at 0°C and maintaining a 20-minute stirring period for each washing; also at 0°C the second batch, BG 2, when assayed showed OD 420 mμ 0.761; cellulose control 3, 0.117; substrate control, 0.024; after a 1-hour incubation at 36°C instead of 3.5 hours as previously.

Example 4: Coupling of Cellulose Carbonate with Trypsin – Cellulose carbonate in 200 mg aliquots was suspended in water (10 ml) and coupled with trypsin (0.0261 gram) in 10 ml water (a) without addition of triethylamine (b) with addition of 20 μl triethylamine and (c) with addition of 200 μl triethylamine. A control experiment was performed in which the conditions of (b) were used except for the replacement of cellulose carbonate by cellulose (200 mg). After stirring at 4°C for 3 hours, the normal procedure of washing was followed but using 0.05M phosphate pH 8.0 instead of acetate.

Each product was finally suspended in 0.5M phosphate buffer, pH 8.0, (10 ml) and an aliquot (0.5 ml) assayed by using N-α-benzoyl-L-arginine ethyl ester hydrochloride as substrate (0.5 ml; 0.0795 gram in phosphate buffer) and incubating at 37°C for 30 minutes. The OD 253 mμ was read on the sixteen times diluted samples after cooling in ice water for 10 minutes and centrifuging. After deduction of a substrate control, the OD 253 mμ values were (a) 0.215 (b) 0.197 (c) 0.792 compared with a cellulose control of 0.026.

Example 5: Coupling of Glucoamylase to Cellulose Carbonate — Cellulose carbonate (100 mg) was stirred with a solution of glucoamylase (e.g., *Aspergillus niger,* 27 units per mg protein, 3.6 mg protein per ml in 0.1M phosphate buffer pH 7.8; 1.0 ml) at 4°C for 3 hours. The solid was filtered off, washed and finally suspended in 0.2M acetate buffer, pH 4.0 (5 ml). A control reaction was carried out in the same way using 1.0 ml of glucoamylase solution and 100 mg cellulose.

Assay for Glucoamylase Activity — Starch solution (1% in 0.2M acetate buffer, pH 4.0; 4.5 ml) was incubated with stirring with the suspension of the solid phase enzyme (0.5 ml) for 1 hour at 45°C. The assay mixture was then cooled and centrifuged and 0.1 ml of the supernatant taken for glucose determination in the glucose oxidase assay for which a standard reference curve was established. The solid phase enzyme preparation caused release of 35 μg of glucose while the control preparation caused release of 17 μg.

Example 6: Coupling of β-Glucosidase with a Diazotized Diaminobenzene Derivative of Cellulose Carbonate — Diazotized diaminobenzene derivative: m-Diaminobenzene (distilled in vacuo) was dissolved in a mixture of dimethylformamide (100 ml), water (100 ml) and triethylamine (2.5 ml). Cellulose carbonate (500 mg) was added slowly to the stirred solution. The mixture was stirred overnight at room temperature and then filtered. The residue was washed with dimethylformamide (100 ml), 0.25 acetate buffer, pH 5.0 (100 ml) and water (200 ml), and then it was dried in vacuo over P_2O_5.

A sample of this material (100 mg) was stirred at 0°C in N hydrochloric acid solution (5.0 ml), sodium nitrite solution (2%, 5.0 ml) was added slowly and the mixture stirred at 0°C for 15 minutes. The solid was then recovered by centrifugation and washed three times by stirring for 15 minutes with 0.1M phosphate buffer, pH 7.8 (10 ml).

Coupling of β-Glucosidase — The solid prepared as above was stirred with β-glucosidase solution (5 mg in 0.1M phosphate buffer, pH 7.8; 20 ml) at 5°C for 18 hours. An ice-cold saturated solution of β-naphthol in saturated sodium acetate solution (5 ml) was added and the mixture stirred for 15 minutes. The solid was then recovered by centrifugation and washed twice as described above. A control preparation was prepared in exactly the same way using cellulose in place of the diazotized diaminobenzene derivative of cellulose carbonate. Both were made up to 5 ml with 0.05M acetate buffer, pH 5.0.

Assay of β-Glucosidase Activity — This was carried out in the normal manner. The test preparation liberated 1.06 μmol of o-nitrophenol in 15 minutes at 37°C while the control liberated 0.05 μmol.

Example 7: Coupling of Glucoamylase with Diazotized Diaminobenzene Derivative of Cellulose Carbonate — The diazotized diaminobenzene derivative (100 mg) was prepared. Glucoamylase solution (1.0 ml) was coupled to this derivative. The washing procedure was carried out and the solid product was finally suspended in 0.2M acetate buffer, pH 4.0 (5 ml). A control experiment was carried out in which cellulose replaced the diazotized derivative. The enzymic activities of the test and control preparations were assayed. The solid phase enzyme liberated 69 μg of glucose per ml in 1 hour while the control liberated 23 μg.

SYNTHETIC POLYMERS AS SUPPORTS

Ethylene Maleic Anhydride (EMA) Polymers

Y. Levin, S. Blumberg, E. Katchalski and L. Goldstein; U.S. Patent 3,650,900; March 21, 1972; assigned to Yeda Research and Development Co., Ltd. have found that papain/EMA type polymeric products can be prepared without loss of the native enzymatic activity. Preparation of papain-polymeric and related compounds according to usual procedure is difficult and that an improved process, which involves (a) inactivating certain portions of the molecule, as by reaction with a metal, e.g., mercury, (b) reaction of the inactivated enzyme with the polymer to give an enzymatically inactivated polymeric product, and (c) thereafter restoring the active sites in the enzyme portion of the polymer-enzyme molecule to yield the enzymatically active polymeric-enzyme product, permits a smoother reaction, greater versatility, and increased yields of desired insoluble product having higher percentages of original native enzyme activity.

Moreover, such an enzyme polymer product can be prepared using relatively crude and highly colored starting enzyme material to give the desired insoluble enzyme-polymer material in substantially colorless form by selective reaction of the polymer with the enzyme in preference to the objectionable contaminating colored impurities.

The process is generally applicable to the preparation of enzyme-polymeric products wherein an enzyme is to be covalently bonded to the polymer molecule. Inasmuch as the enzymatic activity of the starting enzyme is desired to be retained in the final product, it is of course firstly necessary that the bonding of the enzyme to the polymer be through groups which will not inactivate the active sites in the enzyme molecule.

However, when a plurality of reactive groups are present in the enzyme molecule, which are in competition for the reactive groups of the polymer, as in the case of EMA and papain, where not only amino groups but also sulfhydryl groups, among others, can react with the polymer, such competion can and frequently does result in less than optimum activity being imparted to the polymer-enzyme product, due at least partially to competitive reaction between these various reactive groups, including partial inactivation of the end product if the reacting groups constitute a part of the active centers of the enzyme molecule and are necessary for the enzymatic activity, and also because of incidental partial inactivation of activity by destruction of active sites in the enzyme molecule during the course of the reaction by reagent, solvent, or other reaction conditions.

Among the various competitively reactive groups of the various enzyme molecules may be mentioned, besides amino and sulfhydryl, also hydroxyl (including phenolic hydroxyl), carboxyl and imidazolyl. Such groups may be present in inactive portions of the enzyme molecule, as in a cysteine, serine, threonine, histidine, or tyrosine moiety of an enzyme molecule, where the particular moiety in question is not considered essential for enzymatic activity, but they frequently are also present in active portions of the same enzyme molecule, in which event the competitive reactions of such groups result in at least some degree of inactivation during attachment to the polymer molecule.

For the foregoing reasons, it has been found advantageous to inactivate, block, or protect one or more of certain groups in the enzyme molecule prior to reaction with the polymer molecule, then conduct the coupling reaction (by which the enzyme moiety is attached to the polymer in its inactivated, blocked, or protected condition to produce an enzymatically-inactivated enzyme-polymer product, and thereafter to reactivate or unblock the previously inactivated group or groups in the enzyme moiety to provide the desired enzymatically active polymeric product. One of the simplest inactivating or protecting devices has been found to be formation of a metal derivative of the enzyme, for instance the mercury or zinc derivative of an enzyme having at least one sulfhydryl group, e.g., the mercury derivative of papain.

The use of nonoxidizing conditions during formation of such derivatives, as by presence of a mild reducing agent, e.g., cysteine, is advantageous to prevent oxidation of the free sulfhydryl groups in the enzyme during their inactivation. Metal derivatives of other enzymes may also be formed in this manner for inactivation or protection of an active site of the enzyme, e.g., chymopapain, ficin, bromelain, and asclepain, during attachment to the polymer (especially a polymer having carboxyl or anhydride groups) through other groups, which may be of the same type as inactivated, but which are nonessential for activity.

Preferably the metal or other enzyme derivative is also soluble in water and aqueous buffer solutions, but it is only essential that the degree of solubility in water or mixed solvents be sufficient to permit the subsequent coupling reaction. Inactivation and protection does not itself depend on solubility factors. Other inactivating or blocking groups may also be used to produce inactivated or blocked enzyme derivatives, such as pyranyl or tetrahydropyranyl ethers, acrylates, ketals, enamines, enolethers, esters, and the like, according to known procedures for formation of such groups, and their ultimate removal from the enzyme-polymer molecule to restore the enzymatic activity is also according to mildest possible conditions according to known procedure for removal of these groups in other fields, although it has not heretofore been suggested that such inactivation and removal for reactivation would be possible.

The exact nature of the element or group used for inactivating, protecting, or blocking purposes is immaterial, so long as it performs its desired function of inactivation of active sites or groups and so long as it is removable to restore such active sites or groups after incorporation of the enzyme into the polymer molecule without damage to the product. Moreover, complete inactivation of the enzymatic activity of the enzyme and the intermediate protected enzyme-polymeric product is not essential, and in some cases a substantial amount of enzymatic activity remains in the inactivated enzyme and polymeric inactivated enzyme materials. Zinc papain, for instance, may retain up to 20% of the original enzymatic activity, as may the EMA-Zn papain product. After reaction of this protected, blocked, or inactivated enzyme molecule with the polymer to produce the inactivated polymer-enzyme product, e.g., EMA-mercuripapain, in the usual manner and according to usual procedure, the inactivating element or group is also removed according to conventional biochemical procedures, such as the use of a chelating in the case of a metal inactivating agent. When removal of a metal is involved, this is preferably effected under nonoxidizing conditions so as to prevent oxidation of sulfhydryl groups which are being reactivated. The presence of a mild reducing agent, such as sulfhydryl-containing compound, is advantageous in such case, e.g., cysteine, glutathione, methyl mercaptan, thioglycolic acid, mercaptoethanol, or the like.

Nonoxidizing conditions may also be attained by exclusion of atmospheric oxygen by any of various known means, but use of a mild reducing agent is preferred as it ensures against oxidation by oxidation mechanisms which are indigenous to the reaction mixture including even the enzyme itself. In the case of other protecting or inactivating groups, removal is carried out in conventional manner for restoring the integrity of the type of group required for activity of the enzyme moiety, and according to usual mild reaction conditions, temperatures, and relatively neutral pHs as ordinarily employed in the biochemical art.

The end product of the reaction, then, is again the desired insoluble active enzyme-polymer, but ordinarily in higher yields and conversions to desirably active products, and introduction of desirably high amounts of enzyme activity into the polymer molecule is generally much less difficult using the inactivation, coupling, and reactivation process just discussed. As for the polymer in such reaction, it preferably contains carboxyl or anhydride linkages, especially where the enzyme contains an amino, hydroxyl (including phenolic hydroxyl), or sulfhydryl group not essential for its enzymatic activity.

Where the enzyme contains a carboxyl group not essential for activity, the polymer can contain free hydroxyl or amine groups for reaction therewith. The polymer is preferably EMA or an EMA-type polymer, but it can be any of those types previously disclosed for coupling reactions with an enzyme, and in any event it is adapted to effect covalent bonding with the enzyme to produce an enzyme-polymer product either directly or using an activating agent.

Water-insoluble products are produced by reacting the enzyme with a water-insoluble polymer or by causing the reaction product of the enzyme-polymer or inactivated enzyme-polymer to become insoluble either by reaction with a polyfunctional cross-linking agent, such as polyamine or polyol (including glycol), when this is necessary. The reaction product of the enzyme-polymer or inactivated enzyme-polymer product is often insoluble per se because of interaction between the enzyme moiety and additional polymer chains.

If the polymer is precross-linked so as to have a three-dimensional structure or, in some cases, has a sufficiently long linear chain length, the starting polymer is already water-insoluble. Other methods of cross-linking exist and are well-known in the art. Further detailed description follows. Insolubilization via cross-linking can be introduced at any of three stages in the preparation of products of this process.

(1)   The carrier polymer may be cross-linked prior to attachment of the enzyme by any of several procedures well-known in the art of polymer reactions (e.g., incorporation of multifunctional unsaturated monomers during preparation of the polymer or subsequent reaction of the polymer with a few mol percent of multifunctional amines, glycols, etc.).

(2)   Multifunctional amines, glycols, etc., can be added concurrently with the enzyme in the enzyme-attachment or coupling step.

(3)   A multifunctional cross-linking agent may be added to the product after the enzyme has been attached. Such cross-linking agents are added in controllable amounts sufficient to insolubilize the product.

Three-dimensional polymers have an infinite molecular weight. In addition, the enzyme reactant to be attached or coupled to the polymer is commonly multifunctional in itself and thus contributes to the three-dimensional network character of the product. In fact, in many cases, the insolubilization effected in this manner alone is sufficient to impart insoluble characteristics to the product without use of additional cross-linking agents.

When markedly insoluble products are the objective, it is often advantageous to employ copolymers which already contain some cross-linking. Such cross-linked copolymers are known and are obtainable by conducting the polymerization, e.g., the copolymerization of maleic anhydride and hydrocarbon olefin, in the presence of a cross-linking agent, e.g., a compound containing two olefinic double bonds, such as divinylbenzene or vinyl crotonate, poly-1,2-butadiene or alpha, omega-diolefins. The quantity of cross-linking agent will vary with the degree of insolubility desired, but generally will be on the order of from 0.1% to 10% by weight of the total monomer mixture.

Examples 1 through 5: Water-Insoluble EMA-Papain Derivatives — A series of EMA derivatives of papain were prepared and their properties investigated (Tables 1, 2 and 3). The general procedure for preparation of the water-insoluble EMA-papain derivatives listed in Table 1 was as follows:

A solution of the desired amount of papain was prepared by the addition of water to a suspension of crystalline papain in 0.05M acetate buffer, pH 4.5, until a clear solution was obtained. The papain solution was added with stirring to a homogenized suspension of EMA in 0.1M phosphate at pH 7.6. The cross-linking agent hexamethylenediamine was then added, as indicated. The reaction mixture was left to stand overnight in the cold with magnetic stirring. The insoluble EMA-papain derivatives were separated from the reaction mixture and washed by one of the following two procedures:

(a)   Centrifugation and washing with 0.1M phosphate buffer pH 7.5 and then with 0.1M NaCl until activity-free supernatants were obtained, followed by several washings with distilled water. The removal of the salt in the last step caused swelling of the EMA-papain gel.

(b)   Filtration through a sintered glass filter and successive washings, on the filter, with 0.1M phosphate, pH 7.5, followed with 0.1M NaCl.

Finally the salt was removed by suspending the precipitate in distilled water and centrifugation. Method (b) was used only in cases where relatively dense and flaky precipitates were obtained (Nos. 2, 3 and 4 in Table 1).

These EMA-papain suspensions could be filtered only when suspended in 0.1M salt [Procedure (b)]. When an inactivated metal derivative of the polymer papain products is to be tested, e.g., the mercury derivative of EMA-papain, the product to be tested must have the inactivating metal ion removed either prior to or during the assay by addition of an appropriate chelating agent, preferably in a non-oxidizing media, which prevents oxidation of oxidizable groups, e.g., sulfhydryl groups, of the enzyme. In the foregoing systems, these are provided by the EDTA and cysteine.

Assay of Enzymatic Activity — The esterase activity of papain and the EMA-papain derivatives was determined by the pH-stat method at 25°C. The reaction mixture (5 ml) was 0.05M in benzoyl-L-arginine ethyl ester (BAEE), 0.005M in cysteine, and 0.002M in ethylenediaminetetraacetic acid (EDTA). Papain (40 to 100 μg) or EMA-papain derivative (possessing similar esterase activity) was added to the reaction mixture and the rate of hydrolysis followed at pH 6.3 in the case of papain, or at pH 8 to 8.1 in the case of EMA-papain derivatives. NaOH (0.1M) was used as titrant.

The two crystalline papain preparations used in this stody possessed the following esterase activities: commercial 2X crystallized papain, 20 μmols/min/mg protein; commercial 2X crystallized papain, 14 μmols/min/mg protein.

The protease activities of papain and the EMA-papain derivatives were determined by the casein digestion method. The reaction mixture contained 1 ml of a 1% casein solution [in 0.1M tris(hydroxymethyl)aminomethane buffer, pH 7.5], 0.2 ml EDTA 0.02M. Papain (2 to 12 μg) or EMA-papain, possessing similar esterase activity, was added and the casein mixture incubated at 37°C for 20 min. The reaction was stopped by the addition of 3 ml of 5% trichloroacetic acid (TCA), and the optical density determined at 280 mμ.

This convenient spectrophotometric method of analysis involves simply determmining the optical density (OD) at 280 mμ and comparing this figure with the figure obtained from an assay of an equivalent concentration of a known papain or EMA-papain sample.

The protease activity of the EMA-papain derivatives was expressed in terms of mg bound papain, based upon native papain as a standard. The data on the esterase and protease activities of various EMA-papain derivatives is summarized in Tables 2 and 3, respectively.

In the range of carrier to papain ratios studied, the yields of EMA bound papain, as estimated from the esterase activities of the EMA-papain derivatives, were about 50% for the excess carrier preparations (Nos. 1, 2, and 5, Table 2) and somewhat lower (about 35%) for one excess enzyme preparation (No. 4, Table 2). In a preparation in which precross-linked EMA was used (No. 3, Table 2), about 30% of binding was found.

The protease activity of the various EMA-papain derivatives was usually lower by 20 to 40% than the corresponding esterase activity (Table 3), probably due to both steric and electrostatic effects. It should be noted, however, that in all other insolubilized enzyme systems studied, much lower protease to esterase ratios were found. Therefore, unpredictably high protease activities are attained with these immobilized papain derivatives. Lyophilization caused a 10 to 40% drop in enzymic activity (see Tables 2 and 3).

TABLE 1: PREPARATION OF EMA/PAPAIN DERIVATIVES

Example	Carrier to Enzyme Ratio (w/w)	Type of EMA	Degree of Cross-linking with HMD[4], %	Papain in Reaction Mixture	Carrier in Reaction Mixture	Method of Separation and Washing of ppt[3]
1	5:1	Linear EMA[6]	10	210 mg[1] (in 60 ml water)	1,050 mg (in 240 ml 0.125M PO$_4$)	(b)
2	5:1	Linear EMA[6]	5	210 mg[1] (in 60 ml water)	1,050 mg (in 240 ml 0.125M PO$_4$)	(a)
3	5:1	Precross-linked EMA[7]	--	210 mg[1] (in 60 ml water)	1,050 mg (in 240 ml 0.125M PO$_4$)	(a)
4	1:3	Linear EMA[6]	--	210 mg[1] (in 40 ml water)	70 mg (in 40 ml 0.2M PO$_4$)	(b)
5	5:1	Linear EMA[6]	10	220 mg[2]	1,050 mg (in 100 ml 0.2M PO$_4$)	(b)

[1]Worthington, 2X cryst, papain, having an esterase activity towards BAEE[5], pH 6.3 of 20 μmols/min/mg protein.

[2]Worthington, 2X cryst, papain, having an esterase activity towards BAEE[5], pH 6.3 of 14 μmols/min/mg protein.

[3]See text.

[4]Based on weight of carrier; HMD is hexamethylenediamine; EMA is ethylene maleic anhydride copolymer.

[5]BAEE is benzoyl-L-arginine ethyl ester.

[6]MW of about 60,000; DP of about 450 to 500.

[7]Precross-linked EMA[6] is cross-linked with vinyl crotonate.

TABLE 2: ESTERASE ACTIVITY OF THE EMA-PAPAIN DERIVATIVES

Preparation No.	Carrier to Enzyme Ratio (w/w)	Papain in Reaction Mixture, mg	--EMA-Papain, Suspension--		EMA-Papain, Lyophilized Powder		
			Total Esterase Activity[1]	Bound Esterase in Percent of Initial Activity	Total Dry Weight, g	Esterase Activity per mg Powder[1]	Total Esterase Activity[1]
1	5:1	210	99	47	1.7	0.035	59
2	5:1	210	89	40	1.25	0.045	56
3	5:1[2]	210	69	33	1.5	0.041	62
4	1:3	210	71	34	0.21	0.206	43
5	5:1	220	102	46	--	--	--

[1]The enzymic activities are expressed in terms of the number of mg of crystalline papain possessing the same esterase activity.

[2]Cross-linked EMA.

TABLE 3: PROTEASE ACTIVITY OF THE EMA-PAPAIN DERIVATIVES

	Carrier to enzyme ratio (w./w.)	Papain in reaction mixture, mg.	EMA-Papain, lyophilized powder			
			Total dry weight, g.	Protease activity per mg. lyophilized powder [1]	Total protease activity [1]	Bound protease (in percent of initial activity)
No. of prep.:						
1	5:1	210	1.70	0.035	59.5	28.5
2	5:1	210	1.25	0.032	40	19
3	[2]5:1	210	1.50	0.032	48	23
4	1:3	210	0.21	0.090	19	9.5
5	5:1	220	....	....	[3]27	12

[1]The enzymic activities are expressed in terms of the number of
   mg of native crystalline papain, possessing the same protease activity.
[2]Cross-linked EMA.
[3]In aqueous suspension. Sample No. 5 was not lyophilized.

S. Blumberg, E. Katchalski and L. Goldstein; U.S. Patent 3,627,640; December 14, 1971; assigned to Yeda Research and Development Co., Ltd., Israel have also described a method of purifying and decoloring a relatively crude enzyme material, e.g., enzyme or enzyme derivative, by forming a polymer-enzyme product.

Purification, including removal of undesirable colored impurities, from a crude enzyme or enzyme derivative can be advantageously effected by forming a polymeric derivative directly from the crude enzyme or enzyme derivative, as representatively exemplified by papain and mercuripapain. Preferably the polymeric material contains carboxyl or anhydride linkages, especially where the enzyme contains a free amino, hydroxyl (including phenolic hydroxyl), or sulfhydryl group not essential for its enzymatic activity.

Where the enzyme contains a carboxyl group not essential for activity, the polymer can contain free hydroxyl or amine groups for reaction herewith. The polymer is preferably EMA or an EMA-type polymer, but it can be any of those types previously disclosed for coupling reactions with an enzyme and in any event it is adapted to effect covalent bonding with the enzyme to produce an enzyme-polymer product either directly or using an activating agent. The employment of these colored impure (crude) enzyme starting materials for reaction with the polymer in usual aqueous solution or suspension results in selective pickup of the enzyme to the exclusion of impurities and especially pigmented materials, which upon separation remain in the supernatant liquid and can be readily be removed and discarded.

This is especially true when an insoluble polymer-enzyme product is, as conventional, washed with water and the usual buffer solutions, depending upon the exact polymer-enzyme product being treated and the exact pH conditions therefore desired to be maintained. Since enzymes are produced by cell culture processing, including fermentation production from bacteria and animal cell cultures, or from natural materials by extraction, they frequently contain color bodies as impurities besides toxins, pyrogens, and other undesirable contaminants which are removed only with difficulty. Removal of the purified enzyme from such a colored and impure natural milieux by the procedure here provided not only affords a method of purifying the enzyme, but it provides a polymer-enzyme product, either active or inactive (depending upon the starting enzyme, whether native or inactivated) which is substantially devoid of the undesirable colored impurities present in the starting colored enzyme material. Since these colored impurities are of an unknown nature or derivation, selective pickup of the enzyme by the polymer could not be anticipated. Production of substantially colorless insoluble polymer-enzyme product by reaction of the polymer with the enzyme material (including metal and other enzyme derivatives) having colored impurities in combination therewith, and thereafter aqueously washing the insoluble enzyme-polymer product, is among the objects of the process.

Of course, if the enzyme-polymer material is inactivated, as due to use of an inactivated form of the enzyme as the starting colored enzyme material, the relatively pure inactivated enzyme-polymer product will ultimately have to be reactivated before use.

Example 1: Papain-Styrene/Maleic Anhydride Copolymers — Coupling of crystalline papain to an alternating styrene-maleic anhydride (1:1) copolymer, in aqueous buffer medium using the conventional procedure at carrier to enzyme ratios of 1:3 to 5:1, yields insoluble polymer-papain derivatives having up to about 20% of the original enzymic activity.

Employment of mercuri or zinc papain produces the same result with somewhat greater facility and a somewhat higher percentage of initial enzymatic activity in the polymer-papain product.

When relatively crude papain or mercuri or zinc papain is employed, the product after the usual washing with water and aqueous buffer solutions, is characterized by substantial absence of the color originally present in the crude enzyme starting material.

Example 2: Papain-Vinyl Methyl Ether/Maleic Anhydride Copolymers — Coupling of crystalline papain to an alternating vinyl methyl ether/maleic anhydride (1:1) copolymer, in aqueous buffer medium using the conventional procedure at carrier to enzyme ratios of 1:3 to 5:1, yields insoluble polymer-papain derivatives having up to about 50% of the original enzymic activity.

Employment of mercuri or zinc papain produces the same result with somewhat greater facility and a somewhat higher percentage of initial enzymatic activity in the polymer-papain product.

When relatively crude papain or mercuri or zinc papain is employed, the product after the usual washing with water and aqueous buffer solutions is characterized by substantial absence of the color originally present in the crude enzyme starting material.

Example 3: Papain-Divinyl Ether/Maleic Anhydride Cyclocopolymers — Coupling of crystalline papain to an alternating divinyl ether-maleic anhydride cyclocopolymer (having repeating units consisting of adjacent ethylene-maleic anhydride segments which are additionally bonded to each other by an ether linkage), in aqueous buffer medium at carrier to enzyme ratios of 1:3 to 5:1, yields insoluble polymer-papain derivatives having up to about 50% of the original en-

zymic activity. Employment of mercuri or zinc papain produces the same result with somewhat greater facility and a somewhat higher percentage of initial
enzymatic activity in the polymer-papain product.

When relatively crude papain or mercuri or zinc papain is employed, the product
after the usual washing with water and aqueous buffer solutions, is characterized
by substantial absence of the color originally present in the crude enzyme starting material.

EMA Polymers Containing Carboxyl or Carboxylic Anhydride Groups

The procedure of *B.S. Wildi, E.G. Jaworski and T.L. Westman; U.S. Patent
3,616,229; October 26, 1971; assigned to Monsanto Company* involves a polymer-plural enzyme product, wherein the enzymes are covalently bound, comprising one or more proteolytic enzymes, preferably both a neutral protease and an
alkaline protease, and optimally also an additional enzyme, e.g., amylase or lipase,
to further increase the range of enzymatic activity of the product.

The polymer is preferably one containing a free carboxyl or carboxylic anhydride
group adapted to effect covalent bonding with the enzymes either directly or
through activation of a carboxyl group thereof. The polymer may be of relatively low molecular weight and noncross-linked when water-soluble products
are desired, or it may be of higher molecular weight and itself water-insoluble
where a water-insoluble product is desired.

These different types of products will have their own specific fields of application, as in detergents where soluble and in the food and brewing fields where
insoluble. Also, whether soluble or insoluble, they have application in the digestion of organic materials such as sewage or conversion of waste materials and
low-grade protein, e.g., fish, to fertilizer materials. Numerous fields of application for enzymes are already established and the products find ready application
in all such fields, especially because of their stability, and both diverse and wide
range of enzymatic effectiveness. As either soluble or insoluble products, they
are in many cases recoverable for reuse. In any case, they are relatively stable
and long-acting in effect.

EMA is a polymer of ethylene and maleic anhydride. EMA-enzyme is a copolymer of ethylene and maleic anhydride having enzyme covalently bonded thereto.
The product is the same whether the enzyme is reacted directly with an anhydride group of the ethylene-maleic anhydride copolymer or with a carboxyl
group of the ethylene-maleic anhydride copolymer, whether or not using an intermediate activating mechanism for carboxyl groups of the polymer. Anhydride
groups not participating in the reaction by which the product is produced in
aqueous medium are present in the product as carboxyl or carboxylate groups.
Such nonparticipating groups may, however, be converted to amide, imide, ester,
etc., groups, as may be present in EMA-type polymers.

The polymer to which the plurality of enzymes are to be attached contains carboxyl or anhydride linkages, especially where the enzymes contain an amino,
hydroxyl, or sulfhydryl group not essential for their enzymatic activity. Where
an enzyme contains a carboxyl group not essential for activity, the polymer can

contain hydroxyl or amine groups for reaction therewith. The polymer may be EMA or an EMA-type polymer, or be any of those types for coupling or reaction with an enzyme, and in any event it is adapted to couple or react with the enzymes to effect covalent bonding and production of the desired plural enzyme-polymer product.

Since covalent bonding is desired, the carrier polymer is tailored to contain at least one reactive site for each polymer molecule with which the enzymes can react, either directly or indirectly, to produce a covalent bond. This reactive site (or sites) is preferably a carboxyl or carboxylic anhydride group.

The product thus comprises enzymes, including at least one protease and preferably a plurality of proteases, bound covalently through groups which are nonessential for enzymatic activity to a polymer (a) comprising chains of carboxylic acid or carboxylic acid anhydride units, the polymer chains being formed by polymerization of polymerizable acids or anhydrides, or (b) comprising units of carboxylic acid or carboxylic acid anhydride groups separated by carbon chains of at least one and not more than four carbon atoms, the carbon chains being part of a unit which contains a maximum of 18 carbon atoms, the chains usually being formed by copolymerizing a polymerizable acid or anhydride with another copolymerizable monomer, and preferably wherein the starting acid or anhydride and any additional copolymerizable monomer are unsaturated and such polymerization or copolymerization comprises addition type polymerization or copolymerization involving such unsaturation.

Among the polymeric reactants are polyelectrolytes having units of the formula

$$\left[-Z-\overset{O}{\underset{\underset{X}{\overset{\|}{O=C}}}{C}}R_A-(CH_2)_q-\overset{O}{\underset{\underset{Y}{\overset{\|}{C=O}}}{C}}R_B- \right]_n$$

R_A and R_B are selected from the group consisting of hydrogen, halogen (preferably chlorine,) alkyl of one to four carbon atoms (preferably methyl,) cyano, phenyl, or mixtures thereof; provided that not more than one of R_A and R_B is phenyl.

Z is a bivalent radical (preferably alkylene, phenylalkylene, lower-alkoxyalkylene, and lower-aliphatic acyloxyalkylene) comprising a carbon chain having one to four carbon atoms, inclusive, the carbon chain being part of a unit which contains one to 18 carbon atoms inclusive.

q is 0 or 1.

X and Y are selected from hydroxy, $-O$-alkali metal, OR, $-ONH_4$, $-ONHR_3$, $-ONH_2R_2$, $-ONH_3R$, $-NRR'$, $-(Q)_p-W-(NR'R')_x$, where x is 1 to 4 and p is 0 or 1. R is selected from the group consisting of alkyl, phenylalkyl, or phenyl, in each case of one to 18 carbon atoms, where R' is H or R, where Q is oxygen or $-NR'-$, and where W is a bivalent radical preferably selected from lower-alkylene, phenyl, phenylalkyl, phenylalkylphenyl, and alkylphenylalkyl having up to 20 carbon atoms. X and Y taken together can be an oxygen atom, and

at least 1 of X and Y being hydroxyl or X and Y together consti-
tuting an oxygen atom, are preferred.

Many of these polymers are commercially available and others are simple derivatives of commercially available products, which can be readily prepared either
prior to or simultaneously with the enzyme attachment reaction, or produced
as a minor modification of the basic polymer after attachment. Such polymers
containing the above-described EMA-type units are hereinafter referred to as an
EMA-type polymer.

Since enzyme molecules have an extremely high molecular weight, even if the
polymeric units exemplified as usable for attachment of the enzyme occurs only
once in a polymer chain, for example, once in every several hundred units, reaction of the enzyme with this unit will result in an enzyme-polymer product having substantial enzymatic activity and one where the enzyme moiety constitutes
a substantial portion of the molecular weight of the polymeric enzyme product.

If more than one of the exemplified units is present, multiple attachments can
be achieved with increased enzymatic activity of the product. Preferably the
units of the formula given are recurring, n being at least 8. When the units are
recurring, the symbols in the various recurring units do not necessarily stand for
the same thing in all of the recurring units. Moreover, where the units are recurring, some of the X and Y groups may have meanings besides hydroxy or
oxygen. For example, some, but not all, of them may be present in the form
of imide groups, that is, groups in which X and Y together are $-NR-$ or
$-N-W-(NR'R')_x$ where R, W and R' have the values previously assigned.

A preferred type of polymeric material is the polymer of an olefinically unsaturated polycarboxylic acid or derivative with itself or in approximately equimolar
proportions with at least one other monomer copolymerizable therewith. The
polycarboxylic acid derivative can be of the nonvicinal type, including acrylic
acid, acrylic anhydride, methacrylic acid, crotonic acid or their respective derivatives, including partial salts, amides and esters or of the vicinal type, including
maleic, itaconic, citraconic, α,α-dimethylmaleic, α-butylmaleic, α-phenylmaleic,
fumaric, aconitic, α-chloromaleic, α-bromomaleic, α-cyanomaleic acids including
their partial salts, amides and esters. Anhydrides of any of the foregoing acids
are advantageously employed.

Comonomers suitable for use with the above functional monomers include α-
olefins such as ethylene, propylene, isobutylene, 1- or 2-butene, 1-hexene, 1-oc-
tene, 1-decene, 1-dodecene, 1-octadecene, and other vinyl monomers such as
styrene, α-methylstyrene, vinyltoluene, vinyl propionate, vinylamine, vinyl
chloride, vinyl formate, vinyl acetate, vinyl alkyl ethers, e.g., methyl vinyl ether,
alkyl acrylates, alkyl methacrylates, acrylamides and alkylacrylamides or mixtures
of these monomers.

Reactivity of some functional groups in the copolymers resulting from some of
these monomers permits formation of other useful functional groups in the
formed copolymer, including hydroxy, lactone, amine and lactam groups.

Any of the said polybasic acid derivatives may be copolymerized with any of the other monomers described above, and any other monomer which forms a co-polymer dibasic acid derivatives. The polybasic acid derivatives can be copolymers with a plurality of comonomers, in which case the total amount of the comonomers will preferably be about equimolar with respect to the polybasic acid derivatives. Although these copolymers can be prepared by direct polymerization of the various monomers, frequently they are more easily prepared by an after-reaction modification of an existing copolymer.

Copolymers of anhydrides and another monomer can be converted to carboxyl-containing copolymers by reaction with water, and to ammonium, alkali and alkaline earth metal and alkylamine salts thereof by reaction with alkali metal compounds, alkaline earth metal compounds, amines, or ammonia, etc., either prior to, during, or subsequent to enzyme attachment.

Other suitable derivatives of the above polymers include the partial alkyl or other esters and partial amides, alkyl amides, dialkyl amides, phenylalkyl amides or phenyl amides prepared by reacting carboxyl groups on the polymer chain with the selected amines or alkyl or phenylalkyl alcohol as well as amino esters, amino amides, hydroxy amides and hydroxy esters, where the functional groups are separated by lower-alkylene, phenyl, phenylalkyl, phenylalkylphenyl, or alkyl-phenylalkyl, which are prepared in the same manner in each case with due consideration of preservation of enzyme attachment sites as previously stated.

Other aryl groups may be present in place of phenyl groups. Particularly useful derivatives are those in which negatively charged carboxyl groups are partially replaced with amine or amine salt groups. These are formed by reaction of the carboxyls with polyamines such as dimethylaminopropylamine or dialkylamino-alcohols such as dimethylaminoethanol, the former forming an amide linkage with the polymer and the latter an ester linkage. Suitable selection of the above derivatives permit control of several parameters of performance for the enzyme-polymer products of the process.

Representative dibasic acid or anhydride-olefin polymers especially maleic acid or anhydride-olefin polymers, of the foregoing type (EMA-type) are known. Generally, the copolymers are prepared by reacting ethylene or other unsaturated monomer or mixtures thereof, as previously described, with the acid anhydride in the presence of a peroxide catalyst in an aliphatic or aromatic hydrocarbon solvent for the monomers but nonsolvent for the interpolymer formed.

Suitable solvents include benzene, toluene, xylene, chlorinated benzene and the like. While benzoyl peroxide is usually the preferred catalyst, other peroxides such as acetyl peroxide, butyryl peroxide, di-tertiary-butyl peroxide, lauroyl peroxide and the like, or any of the numerous azocatalysts, are satisfactory since they are soluble in organic solvents. The copolymer preferably contains substantially equimolar quantities of the olefin residue and the anhydride residue.

Generally, it will have a degree of polymerization of 8 to 10,000, preferably about 100 to 5,000, and a molecular weight of about 1,000 to 1,000,000, preferably about 10,000 to 500,000. The properties of the polymer, such as molecular weight, for example, are regulated by proper choice of the catalyst and control of one or more of the variables such as ratio of reactants, temperature, and catalyst concentration or the addition of regulating chain transfer agents, such as diisopropylbenzene, propionic acid, alkyl aldehydes, or the like. The product is obtained in solid form and is recovered by filtration, centrifugation or the like. Removal of any residual or adherent solvent can be effected by evaporation using moderate heating. Numerous of these polymers are commercially available. Particularly valuable copolymers are those derived from ethylene and maleic anhydride in approximately equimolar proportions. The product is commercially available.

The maleic anhydride copolymers thus obtained have repeating anhydride linkages in the molecule, which are readily hydrolyzed by water to yield the acid form of the copolymer, rate of hydrolysis being proportional to temperature. In view of the fact that attachment reactions of the process are carried out in aqueous solutions or suspensions, or using water-solvent mixtures, the product of the reaction or coupling of the enzyme to EMA has carboxyl or carboxylate groups attached to its chains adjacent the attached enzyme instead of anhydride groups, due to hydrolysis of the anhydride groups, which do not react with the enzymes, during the reaction. The same is true of nonreacting anhydride groups present in other polymers, such as EMA-type polymers, which hydrolyze to carboxylate groups during the reaction.

The term water-insoluble means that the product concerned does not dissolve in water or aqueous solutions, even though it may have such characteristics as a high degree of swelling due to solvation by water, even to the extent of existence in a gel form. Water-insoluble products can be separated by methods including filtration, centrifugation, or sedimentation. Such characteristics are imparted by cross-linking.

Water-insoluble plural enzyme-polymer products are produced by reacting the enzymes with a water-insoluble polymer or by causing the reaction product of the enzymes and polymer to become insoluble either by reaction with a polyfunctional cross-linking agent, such as a polyamine or polyol (including glycol), when this is necessary. The plural enzyme-polymer product is frequently at least in part insoluble per se because of interaction between the enzyme moiety and additional polymer chains. If the polymer is precross-linked so as to have a three-dimensional structure or, in some cases, has a sufficiently long linear chain length, the starting polymer is already water-insoluble.

Insolubilization via cross-linking can be introduced at any of three stages in the preparation of products.

(1) The carrier polymer may be cross-linked prior to attachment of the enzyme by any of several procedures (e.g., incroporation of multifunctional unsaturated monomers during preparation of the polymer or subsequent reaction of the polymer with a few mol percent of multifunctional amines, glycols, etc.).

(2) Multifunctional amines, glycols, etc., can be added concurrently with the enzyme in the enzyme-attachment step.

(3) A multifunctional cross-linking agent may be added to the product after the enzyme has been attached. Such cross-linking agents are added in controllable amounts sufficient to insolubilize the product.

In addition, the enzyme reactant to be attached, e.g., coupled, to the polymer is commonly multifunctional in itself and thus contributes to the three-dimensional network character of the product. In fact, in many cases, the insolubilization effected in this manner alone is sufficient to impart insoluble characteristics to the product without use of additional cross-linking agents.

When markedly insoluble products are the objective, it is often advantageous to employ copolymers which already contain some cross-linking. Such cross-linked copolymers are known and are obtainable by conducting the polymerization, e.g., the copolymerization of maleic anhydride and hydrocarbon olefin, in the presence of a cross-linking agent, e.g., a compound containing two olefinic double bonds, such as divinylbenzene or vinyl crotonate, poly-1,2-butadiene or alpha, omega-diolefins. The quantity of cross-linking agent will vary with the degree of insolubility desired, but generally will be on the order of from 0.1% to 10% by weight of the total monomer mixture.

The general procedure is as follows: Crude *B. subtilis* enzyme mixture is suspended in cold distilled water and stirred magnetically for one hour at 4°C. The resulting mixture is then centrifuged at 8,000 rpm for 10 minutes to remove suspended and inactive solids. The supernatant is separated and made 0.065M in calcium ion by the addition of 1M $Ca(OAc)_2$ and the solution is then stirred for 30 minutes in the cold (4°C). The mixture is then centrifuged at 8,000 rpm for 10 minutes to remove precipitated and inactive solids.

To the clarified supernatant there is added, with stirring, cold 0.05M Veronal buffer, pH 7.8. While the above solutions are being prepared an appropriate quantity of EMA is dissolved in dimethyl sulfoxide. This solution is added dropwise to the stirred, cold enzyme solution and the mixture is then stirred overnight at 4°C. The mixture is then centrifuged at 8,000 rpm for 10 minutes and the solid product is collected. The solid adduct is washed using twice its volume of cold, distilled water, with stirring and centrifugation. The adducts were washed in this manner 15 times and the product was then isolated by lyophilization.

The yield of insoluble products is advantageously achieved, when desired, by performing the reaction in the presence of a cross-linking agent such as hexamethylenediamine, e.g., at a 1 to 2% concentration relative to the amount of polymer employed.

Example 1: Plural Protease/EMA Insoluble and Soluble Adducts — *B. subtilis* alkaline and neutral proteases (200 mg) are dissolved in 50 ml cold 0.1M in phosphate and 0.01M in calcium acetate, pH 7.5, and this solution is then added to a cold, homogenized mixture of EMA-21 (100 mg) suspended in 50 ml 0.1M phosphate, pH 7.5. The mixture is stirred overnight in the cold (4°C) and the insoluble material is separated from the supernatant by centrifugation. After washing the solids five times with cold 0.1M NaCl and twice with water, the material is lyophilized to yield a solid which possesses 38% of neutral protease activity and 52% of the original alkaline protease activity.

The supernatant solution is dialyzed against cold, distilled water and then lyophilized to yield a soluble solid which possesses 47% of the neutral protease activity and 59% of the original alkaline protease activity.

*Example 2: Neutral Protease/Lipase/Cellulase-Styrene/Maleic Anhydride Co-
polymers* — Coupling of a mixture of neutral protease, lipase and cellulase to
an alternating styrene-maleic anhydride (1:1) copolymer, in aqueous buffer medium using the conventional procedure of Example 1 at carrier to enzyme ratios 1:3 to 2:1, yields insoluble polymer-neutral protease/lipase/cellulase derivatives having up to about 20% of the original enzymatic activities. Employment
of 1% hexamethylenediamine increases the amount of insoluble cross-linked
product.

Organosilane Derivatives of EMA

According to *R.E. Miller; U.S. Patent 3,715,278; February 6, 1973; assigned to
Monsanto Company* enzyme-polymer products are covalently bound to siliceous
materials by siliation of the siliceous material to provide functional groups which
react with functional groups in the polymer-enzyme molecule. Alternatively,
the polymer is first reacted with the reactive groups of the organosilated siliceous
material and the enzyme subsequently reacted with the product of this reaction.
The product in either case is an insoluble enzymatically-active siliceous material
in which the enzyme is covalently bound to the polymer molecule which is in
turn covalently bound to the surface of the organosilated siliceous material.

It is possible to realize both the potential for stabilizing and modifying enzyme
reactions by surrounding the enzyme with a polyelectrolyte environment and
to obtain a particle size and shape consistent with packing columns so as to
achieve more useful flow rates. In addition, any of several low-priced siliceous
materials are suitable carriers for the catalytically active product.

To achieve attachment to the carriers, the carriers are first modified by covalently
binding the polymeric ionic material (polyanions, polycations, polyampholytes,
etc.) to the surface by means of a coupling agent followed by attachment of the
bio-active enzyme catalyst to the polymeric surface. In this manner better flow-
through rates and greater thermal and autolytic stability can be achieved. Additionally, the reactions can be carried out at optimum pH conditions and taking
full advantage of substrate specificity.

The polymer is preferably one containing a free carboxyl or carboxylic anhydride group adapted to effect covalent bonding with the enzyme or enzymes
either directly or through activation of a carboxyl group thereof. The polymer
may be of relatively low molecular weight and noncross-linked or it may be of
higher molecular weight and be itself water-insoluble.

Enzyme-polymer derivatives can be prepared by reacting a crystalline or crude
enzyme or enzyme mixtures with the polymer in solution, resulting in formation of a polymeric product in which the enzymes are covalently bound, and
which, in turn, is subsequently attached to an organosilane-treated siliceous material. In the alternative, the polymer is attached to the siliceous material via
an organosilane coupling agent first and the enzyme then attached to the polymer.

The reaction of the polymer with a plurality of enzymes can be carried out step-
wise, one enzyme at a time, with or without intermediate isolation, or with several enzymes at once. The latter procedure is preferred for reasons of time, con-

venience and economy when more than one enzyme or the material is desired. When an anhydride or carboxyl is present in the polymer, e.g., an EMA-type polymer, covalent bonding of the enzyme to the polymer may be effected directly through reaction or coupling with an anhydride group or with carboxyl group using an activating agent. The product is the same in both cases. The pH range for the reaction depends on the enzymes employed and their stability ranges. It is usually about 5 to 9.5, preferably about 6 to 8, but adjustment must be made for individual cases.

Isolation and purification is generally effected according to normal biochemical procedures, and by the general procedure of the examples which follow. Since covalent bonding of the enzyme to the polymer is desired, the reaction is ordinarily carried out at low temperatures and at relatively neutral pHs, in water or dilute aqueous buffer as solvent.

When carried out in this manner, the results are production of the desired active enzyme-polymer derivative, but degree of activity imparted to the polymeric product is sometimes lower than desired, possibly due to partial inactivation of the enzyme during the process. Resort may frequently advantageously be had to employ of a mixed solvent system, using a solvent in which the enzyme is at least partially soluble, usually in an amount up to about 50% by volume.

Dimethyl sulfoxide (DMSO) is especially suitable as solvent together with water or aqueous buffer solution in a mixed solvent system. Using such a mixed solvnet system, the desired active enzyme-polymer is ordinarily obtained in higher yields and conversions to desirably active composite, and introduction of desirably high amounts of enzyme activity into the polymer molecule is generally less difficult.

As far as the polymer in such reaction, it preferably contains carboxyl or anhydride linkages, especially where the enzyme contains an amino, hydroxyl (including phenolic hydroxyl), or sulfhydryl group not essential for its enzymatic activity. Where the enzyme contains a carboxyl group not essential for activity, the polymer can contain free hydroxyl or amine groups for reaction therewith.

The polymer is preferably EMA or an EMA-type polymer, but it can be any of those types previously disclosed for coupling or reaction with an enzyme, and in any event it is adapted to effect covalent bonding with the enzyme to produce an enzyme-polymer product either directly or indirectly using an activating agent. Inasmuch as the enzymatic activity of the starting enzyme is desired to be retained in the final product, it is of course firstly necessary that bonding of the enzyme to the polymer be through a group which will not result in inactivation of an active site in the enzyme molecule.

Among the various reactive groups of enzyme molecules may be mentioned, besides amino and sulfhydryl, also hydroxyl (including phenolic hydroxyl), carboxyl and imidazolyl. Such groups are present in free or unbound form in inactive portions of enzyme molecules, as in a lysine, cysteine, serine, threonine, histidine, or tyrosine moiety of an enzyme molecule, where the particular moiety in question is not considered essential for enzymatic activity, either catalytic in nature or for substrate binding. Therefore, attachment to the polymer molecule is through reaction of the polymer with such groups so as to avoid inactivation of the enzymes during attachment to the polymer molecule. Generally, the linkage is an amide, imides, ester, thioester, or disulfide group, such as formed by the carboxyl or anhydride of the polymer with an amine or hydroxyl group in a nonessential moiety of the enzyme protein chain. Amides are conveniently formed by reacting pendant amino groups of the enzyme with carboxylic anhydride groups on the carrier polymer in water, in aqueous buffer media, or in mixed solvents.

Amides, imides and esters are readily formed by activating carboxyl groups of the polymer, or alternatively pendant carboxyls of the enzyme, and reacting them with respective hydroxyl, amine or mercaptan groups on the other reactant. Such activation may be effected using various carbodiimides, carbodiimidazoles, Woodward's or Sheehan's reagent, or the like, to form highly active intermediate capable of reacting with other groups mentioned above under mild conditions, the latter favoring retention of enzymatic activity.

The polymer selected for such reaction can therefore be said to be adapted to couple or react with the plurality of enzymes, either directly or indirectly through use of an activating agent, as already indicated, and in any event covalent bonding with the enzymes. The attachment procedures given are conducted by techniques adapted to include any requisite protection for the enzyme, which may include a reversible blocking of the enzymatically active site or sites, as for example in the case of papain, where mercuripapain or zinc papain may be employed as an intermediate for reaction with the polymer in order to effect greater yields upon attachment, the protecting atoms being removed subsequent to the attachment reaction.

The polymer to which the plurality of enzymes are to be attached contains carboxyl or anhydride linkages, especially where the enzymes contain an amino, hydroxyl, or sulfhydryl group not essential for their enzymatic activity. Where an enzyme contains a carboxyl group not essential for activity, the polymer can contain hydroxyl or amine groups for reaction therewith. The polymer may be EMA or an EMA-type polymer, or be any of those types for coupling or reaction with an enzyme, and it is adapted to couple or react with the enzymes to effect covalent bonding and production of the desired enzyme-polymer composite. The polymer is further described in U.S. Patent 3,616,229 (see page 42).

Since covalent bonding is desired, the polymer is tailored to contain at least one reactive site for each polymer molecule with which the enzymes can react, either directly or indirectly, to produce a covalent bond. This reactive site (or sites) is preferably a carboxyl or carboxylic anhydride group.

In addition, the enzyme reactant to be attached, e.g., coupled to the polymer, is commonly multifunctional in itself and thus contributes to the three-dimensional network character of the product. In fact, in many cases, the insolubilization effected in this manner alone is sufficient to impart insoluble characteristics to the enzyme-polymer product without use of additional cross-linking agents.

Suitable agents for coupling the enzyme-polymer to the siliceous material such as glass beads or fibers, diatomaceous earth, sand, clays, asbestos, and the like, are substituted organosilanes which can be represented by the formula

$$X_a'—\underset{\underset{Y_b'}{|}}{Si}—[R_n'—Z']_c$$

where X' is a hydrolyzable group capable of reacting with a hydroxyl group, Y' is hydrogen or monovalent hydrocarbon group, R is alkylene group having from one to about 20 carbon atoms, Z' is a functional group capable of reacting with the aforementioned polymer, n is an integer having a value of 0 to 1, a is an integer having a value of 0 to 2, inclusive, c is an integer having a value of 1 to 3, inclusive, and the sum of a + b + c equals 4.

Examples of suitable X' groups include halo, hydroxy, alkoxy, cycloalkoxy, aryloxy, alkoxy-substituted alkoxy such as β-methoxyethoxy or the like, alkoxycarbonyl, aryloxycarbonyl, alkyl carboxylate, and aryl carboxylate groups, preferably having eight or less carbon atoms.

Examples of suitable Y' groups in the above formula are hydrogen, methyl, ethyl, vinyl, isobutyl, and other hydrocarbyl groups, preferably having ten or less carbon atoms.

The R' group in the above formula can be any alkylene group having up to about 20 carbon atoms, and preferably from about two to about 18 carbon atoms. Examples of such groups are ethylene, the propylenes, the butylenes, the decylenes, the undecylenes, the octadecylenes, and the like.

The Z' group can be any functional group capable of reacting with the aforesaid polymer. Examples of such groups are amino, primary and secondary amido, epoxy, isocyanato, hydroxy, alkoxycarbonyl, aryloxy carbonyl, vinyl, allyl, and halo such as chloro and bromo groups, and the like.

Particularly preferred organosilane coupling agents for the purposes and of this process are the ω-aminoalkyl- and aminoaryltrialkoxysilanes such as γ-aminopropyltrimethoxysilane, aminophenyltriethoxysilane, and the like.

The siliceous materials that can be utilized include silica in the form of sand, fibers, cloth, or the like; glass in the form of cloth, fibers, matting, or the like, diatomaceous earth, asbestos, also silicates such as wollastonite, fosterite, feldspar, mullite, various clays including bentonite, kaolin, etc. The principal requirement of the siliceous material is that surface hydroxyl groups be present which are reactive with the hydrolyzable groups on the organosilane coupling agent.

The siliceous material can be treated with the organosilane coupling agent in any convenient manner by contacting the former with the latter to obtain the desired bonding. Usually the organosilane is dissolved in an inert solvent such as toluene, xylene, or the like, and the resulting solution is then applied to the siliceous material. Aqueous solutions of the silane can also be used.

The amount of coupling agent employed is dependent upon the nature and surface area of the siliceous material, and also, of course, on the particular polymer to be ultimately attached to the material. Usually at least about 0.01% by weight of the coupling agent, based on the weight of the siliceous material, is desired. Amounts in the range from about 0.25 to about 2.0% by weight are preferred.

Example 1: Enzyme Bonding on Calcium Silicate — Calcium silicate (5.0 grams), coupled with 0.25 weight percent γ-aminopropyltrimethoxysilane and *B. subtilis* enzyme mixture (1.0 gram) are slurried in 20 milliliters of a dimethyl sulfoxide (DMSO) solution of EMA (DMSO to EMA w/w ratio 5:1). The obtained slurry is stirred until gelation which takes place after about 8 minutes.

Thereafter the produced gel is blended, in a laboratory blender, with crushed ice, water (150 ml) and n-propanol (75 ml) for 45 seconds, water (300 ml) added thereto, and the resulting mixture stirred for 72 hours at 4°C. The mixture is then centrifuged, and the recovered solids lyophilized. About 4.3 grams of a solid product having a protease activity of about 1,000 units per gram is obtained.

Example 2: Enzyme Bonding on Silica — In a manner similar to Example 1 but using particulate silica, as the siliceous material, treating the material with γ-aminopropyltrimethoxysilane, and contacting the treated material with a *B. subtilis* enzyme mixture in a dimethyl sulfoxide solution of EMA an insoluble, enzymatically-active composite is obtained.

Example 3: Enzyme Bonding on Wollastonite — In a manner similar to Example 1 but using γ-aminopropyltrimethoxysilane and treating wollastonite therewith, and thereafter contacted the treated wollastonite with a *B. subtilis* enzyme mixture in a dimethyl sulfoxide solution of EMA an insoluble, enzymatically active composite is obtained.

N-Ethyl-5-Phenylisooxazolium-3-Sulfonate Ester of EMA

R.P. Patel; U.S. Patent 3,691,016; September 12, 1972; assigned to Monsanto Company has prepared an insoluble enzyme by dissolving a water-soluble enzyme containing both carboxyl and primary amine moieties in water, adding a 3-unsubstituted isooxazolium salt in a minor molar amount with respect to the carboxyl moiety, reacting the salt and a carboxyl moiety to obtain an activated ester, condensing the activated ester with an amine moiety to produce a water-insoluble enzyme product of increased molecular weight which contains amide linkages.

The carboxyl values utilized in the formation of the activated ester are supplied from either one of two sources, or a combination thereof. The first source being the enzyme which is being rendered insoluble, and the second source being a polyacidic polymer. Typical polycarboxyl polymers include (1) copolymers of a hydrocarbon olefin and a monomer selected from the group consisting of unsaturated polycarboxylic acids, their anhydrides, and salts as well as polymers which are polymerized ethylenically unsaturated acids, (2) homopolymers of polymerizable acids, e.g., polyacrylic and (3) carboxymethylcellulose polymers.

Example 1: A copolymer of ethylene and maleic anhydride (EMA) (mol ratio approximately 1:1) having a molecular weight of about 2,000 was hydrolyzed and converted to the half sodium salt to insure water solubility. 600 mg of EMA salt were dissolved in 10 ml of water. To the solution was added 10 ml of a 2% aqueous solution of N-ethyl-5-phenylisooxazolium-3-sulfonate (Woodward's Reagent K). The mixture was stirred for a few minutes while the temperature was maintained at 4°C. 100 mg of chymotrypsin (CHT) dissolved in water (total volume 10 ml) was added to the reaction mixture.

The stirring was continued for 16 hours until a white precipitate (condensed product) separated out. The precipitate was centrifuged, washed once with cold 0.1M sodium chloride and twice with cold distilled water. The samples were dried at room temperature under reduced pressure and nitrogen determined by the Kjeldahl method, which gives data necessary to calculate the amount of bound protein in the polymer. The chymotrypsin activity of the samples was determined by a titrimetric method, using N-acetyltyrosine ethyl ester as substrate in a 0.01M phosphate buffer at pH 7.5. NaOH solution (0.1N) was used as a titrant. The insoluble enzyme was obtained in a yield of 81%, having a protein content of 76% and a retained activity of 56%.

Examples 2 through 21: Additional insoluble enzyme compositions were prepared using the foregoing procedure but with varying proportions of reactants. The results are summarized in Table 1.

Examples 22 through 31: Another series of experiments were performed using the test conditions of Example 1 except that acetic acid was used to precipitate the resultant insoluble enzyme. Precipitation occurred at about a pH of 4. The results are tabulated in Table 2.

Similar results are obtained when other enzymes such as trypsin, urease are employed. Similar results are also obtained when other ester-forming compounds such as the carbodiimides and the carbodiimidazoles are employed.

TABLE 1: COMPOSITION AND ACTIVITY OF CHYMOTRYPSIN-EMA POLYMERS[1]

Example	EMA/CHT, mg.	Woodward's reagent, g./1 g. EMA	Yield, mg.	Bound protein, percent	Activity retained, percent [2]
2	100:100	0.372	122	61	22
3	600:100		146	55	37
4	100:400		360	75	11
5	100:100	0.303	97	59	22
6	600:100		102	52	48
7	100:400		261	70	7
8	100:100	0.251	98	58	23
9	600:100		104	53	52
10	100:400		269	69	6
11	100:100	0.202	130	61	25
12	600:100		135	55	44
13	100:400		251	69	10
14	100:100	0.186 (0.188)	122 (152)	61 (60)	22 (26)
15	600:100		146 (170)	55 (54)	37 (55)
16	100:400		360 (307)	75 (73)	11 (8)
17	100:100	0.145	158	64	20
18	600:100		186	54	46
19	100:400		257	71	9
20	100:100	0.093	121	60	9
21	600:100		143	58	22

[1] Water system, 2 hour-reaction time and precipitation with conc. HCl to pH 3.
[2] All activities determined at pH 7.5.

TABLE 2: CHYMOTRYPSIN-EMA POLYMERS FROM ACETIC ACID

Example	EMA/CHT, mg.	Woodward's reagent, g./1 g. EMA	Yield, mg.	Bound protein, percent	Activity retained, percent
22	100:100	[1] 1. 480	63	72	40
23		[1] 0. 740	60	81	41
24		0. 372		83	66
25		0. 320	104	70	81
26		0. 240	88	67	88
27		0. 186	[2] 73	77	78
28		0. 093	84	77	68
29		0. 046	82	82	78
30		0. 023	111	76	66
31		0. 006	109	76	72

[1] G./0.8 g. EMA.
[2] Experiment was carried out in phosphate buffer 0.01 M, pH 7.5.

Example 32: One gram of 15.0% polyacrylic acid solution was dissolved in 5 ml of water. To this 1N NaOH solution was added until pH 7 was attained. Woodward's reagent (1.5 grams), dissolved in about 10 ml of water, was added to the above solution, was cooled, and stirred for about half an hour, α-Chymotrypsin (0.150 gram) dissolved in 10 ml of water, was added to the reaction mixture and was stirred over a weekend (an overnight period is sufficient). The white precipitate which separated out was centrifuged, washed three times with distilled water, and was dried at room temperature under reduced pressure.

The yield was 0.250 gram (77% enzyme bound). Nitrogen analysis by Kjeldahl's method showed 7.1% ≡ 46 mg of enzyme/100 mg of polymer. Enzyme activity retained was 3.5%.

Examples 33 through 36: Other experiments employing different ratios of the reactants above were performed. The results are tabulated in Table 3, in each example 150 mg of polyacrylic acid was reacted.

TABLE 3: CHYMOTRYPSIN-POLYACRYLIC ACID POLYMERS

Example	Chymotrypsin, mg.	Woodward's reagent, mg.	Percent activity retained
33	150	500	7. 1
34	300	500	26. 0
35	300	250	57. 0
36	300	10	76. 0

Comparable results are obtained when other acid polymers including polyglutamic acid and polyamic acid polymers are employed. The infrared spectrum of polyacryl-α-chymotrypsin was studied, with reference to the spectra of polyacrylic acid and α-chymotrypsin. α-Chymotrypsin showed NH band at 3μ, amide I band at 6.1μ, and amide II band at 6.6μ.

Polyacrylic acid's main bands were carboxylic OH at 3.2μ and carboxylic carbonyl band at 5.9μ. The infrared spectrum of polyacryl-α-chymotrypsin incorporated characterization of both the parent compounds showing broad bands for NH, OH at 3μ, and carbonyl at 5.8μ and amide I and amide II bands at 6.1μ and 6.6μ respectively. The Σ values of these bands evidenced a high degree of condensation between polyacrylic acid and α-chymotrypsin.

Example 37: Carboxymethylcellulose-α-chymotrypsin was prepared in the same way as mentioned above from carboxymethylcellulose sodium salt (0.200 gram), Woodward's Reagent (0.3 gram), and α-chymotrypsin (0.2 gram). The reaction was carried at 5°C in 0.2M phosphate buffer, pH 7.5. The compound was worked up by the same process as before. Yield was 0.200 gram.

Example 38: The procedure of Example 1 is followed in all essential details with the exception that a mixture of dinitrophenol and 1-cyclohexyl-3-(2-morpholino-ethyl)carbodiimide methyl-p-toluenesulfonate is substituted for N-ethyl-5-phenyl isooxazolium-3-sulfonate to obtain chymotrypsin insoluble form.

Example 39: A strip 2 x 1 inches of cotton gauze was placed in a solution of 0.567 gram of NaOH in 20 ml of water, and the solution was heated at 85°C for an hour and then cooled to room temperature (about 25°C). The gauze was then soaked for 0.75 hour in a solution of 0.033 gram Woodward's Reagent K in 10 ml water, after which it was removed and soaked for an hour in a solution of 0.010 gram chymotrypsin in 5 ml water, both at room temperature.

Finally, the gauze was washed with 0.1M aqueous NaCl and then twice with water, and blotted until the strip was no longer dripping wet. The wash water was free of chymotrypsin, showing that any present was insolubilized.

To measure the activity of the enzymatically active fabric prepared as stated, plates were prepared of 2 grams gelatin in 30 ml water. The treated gauze strip was cut in half. One was placed on a gelatin plate immediately, and the other was left to dry under vacuum.

As controls, gauze strips soaked in aqueous chymotrypsin (ChT) solution and containing respectively about 1.6 mg, 2.2 mg, and 4.5 mg. ChT were placed in other gelatin plates, and gauze strips soaked respectively in NaOH and in NaOH and Woodward's Reagent solutions as described above were placed on other plates.

Later, the vacuum-dried treated gauze was placed on another gelatin plate. The plates were each observed for gelatin liquefaction, which is produced by chymotrypsin activity, with results as tabulated below, showing activity for the damp treated gauze equivalent to that of the gauze carrying 1.6 to 2.2 mg of chymotrypsin, as measured by amount of liquid formed.

	Liquidity				
Elapsed time, hrs	1	2	4	13	39
Sample:					
Insolubilized ChT gauze	None	Some	Great	Extremely	
1.6 mg. soluble ChT	Slight	do	do	do	
2.2 mg. soluble ChT	do	do	do	do	
4.5 mg. soluble ChT	do	Large	Large	do	
NaOH	None	None	None	None	None.
NaOH plus Woodward's reagent	do	do	do	do	Do.
Vacuum dried gauze				(Applied)	Some.

Acrylamide-EMA Polymers

In the methods of *J. Dieter, G. Wolfgang and B.H. Ulrich; U.S. Patent 3,775,253; November 27, 1973; assigned to Boehringer Mannheim GmbH, Germany,* water insoluble biologically active proteins are bound on copolymers consisting of (a) acrylamide (b) ethylene-maleic acid or its anhydride and/or (c) maleic acid and/or its anhydride and (d) N,N'-methylenebisacrylamide or ethylene diacrylate in a weight ratio a:b:c:d of 3:0.5-1.5:0.05-4.0:0.075-0.9, the weight ratio a:(b + c) being not more than 4, to provide compositions which are outstanding in their ability to yield up the bound protein, and with which precise substrate reactions can be carried out.

It is known that insoluble proteins can be formed by condensation, coupling and polymerization reactions with the protein itself. The proteins obtained in this manner are products of uncontrollable reactions; in other words, products with a definite constitution cannot be obtained.

A large part of the protein becomes denatured by the severe reaction conditions of the polymerization and of the coupling with, for example, bis-diazo compounds and by the action of the components used for the condensation, such as formaldehyde, ethyl chloroformate and the like, and thus is biologically inactivated. For most fields of use of insoluble proteins, such denatured material is useless.

As insoluble, biologically-active proteins, only those proteins which are fixed on carrier materials are of importance since they are materials of definite constitution which can generally be used universally as enzymes, inhibitors, antigens or antibodies or as proteins.

The binding of the proteins to the carrier materials can take place in a heteropolar or homopolar manner. The heteropolar bound proteins are of lesser interest since the proteins can be lost by elution from the carrier materials, in amounts depending upon the ion concentration and pH value of solutions with which they are contacted. The following methods have been used for fixing proteins to insoluble carriers by means of reactive groups:

> (1) Reaction of cellulose with p-nitrobenzyl chloride to give the corresponding cellulose derivative, which is reduced to the corresponding amino derivative and diazotized and then coupled with proteins;
>
> (2) The reactive groups described in (1) are replaced by m-aminobenzyloxymethyl radicals;
>
> (3) Polyamino-polystyrene, which is a completely synthetic carrier, is reacted with proteins;
>
> (4) d,l-p-aminophenylalanine, d,l-leucine and similar amino acids are polycondensed to give a synthetic carrier, the diazotized amino groups of which are coupled with proteins;
>
> (5) By means of methods known from peptide chemistry, carboxymethylcellulose is reacted with proteins using dicyclohexyl carbodiimide as condensation agent;

(6) From carboxymethylcellulose there is produced, via the hydrazide, the corresponding azide which is reacted with proteins;

(7) Proteins have been bound with bromoacetylcellulose;

(8) From isothiocyanate derivatives of the dextran gel (Sephadex) cellulose, there have been produced water-insoluble enzyme compounds with proteins;

(9) Under the conditions of acrylamide polymerization, proteins can be fixed in readily swellable material. The enzyme is thereby bound radically on the chains;

(10) Furthermore, proteins have also been bound to ethylene-maleic acid anhydride (EMA);

(11) Proteins have also been bound to styrene-maleic anhydride copolymers.

The previously known methods are not satisfactory and are not of general applicability for proteins. Thus, less than 50% of protein is bound by methods (1), (2) and (4), and a greater part of the protein loses its biological activity. Method (3) has proven to be even less suitable. Method (5) can only be carried out in a nonaqueous medium and, therefore, can only be used for some antibody and low molecular weight inhibitors.

According to method (6), only about 10% of the protein used is bound, only 10 to 43% of which retains its enzymatic activity. Method (7) gives a high yield of bound protein (80 to 90%) but for many proteins, the reaction conditions used result in a loss of biological activity. In the case of method (8), only a very small percentage of the protein used is actively bound to the gel.

According to method (9) only 18 to 20% of the protein used is bound, only a few percent of which can evolve an activity against substrates. High molecular weight substrates, such as hemoglobin, only reacted to an extent of 1 to 2%. Method (10) gives the most satisfactory results. It is used for antigens, inhibitors and enzymes. The binding reaction takes place under mild conditions in an aqueous medium; about 50 to 60% of the protein can be fixed under optimum conditions.

However, the starting material used is very nonuniform. It contains lower molecular weight fractions, some of which are soluble, and higher molecular weight fractions, which are insoluble, which cannot be separated by sieving or by other means. Therefore, only up to 60% of the protein is bound to the insoluble carrier, the remainder being lost as soluble material. However, the bound protein is still not satisfactory.

In the binding of the protein to the carrier (EMA), the protein fulfills the function of a comonomer; it can be bound not only at one position but also at several positions, i.e., it may be cross-linked. Depending upon the protein concentration, more or less readily swellable protein-carrier materials are obtained. The protein is thereby present in a network in covalent bound form so that diffusion plays an important part in the case of the use of this carrier and it is only partially accessible to a substrate. This disadvantage is not even overcome by the addition of cross-linking agents, such as hexamethylene-diamine or the like. Therefore, only about 50% of the bound protein can be reached by low molecular weight substrates, so that the total activity thereof does not amount to more than 30%. In the case of high molecular weight substrates, an activity of only 10% or less has been found.

An additional disadvantage of this material is that it is of varyingly difficult filterability, i.e., it can be flocculent, which differs from one protein to another, and is also dependent upon the protein concentration. Thus, there is no assurance that the bound protein produced therewith can be packed into columns.

Styrene-maleic anhydride polymers used according to method (11) can be used for the removal of proteins, i.e., in the purification of drinking water and the like. These carriers have, similarly to the polyamino-styrenes, a decidedly lyophilic character. Proteins cannot be bound to their structure without loss of activity and, when such polymers are bound with proteins, they cannot be lyophilized.

The functional groups of the known carriers which are not bound to protein carry, in the appropriate pH ranges, positively charged NH_2-groups in the case of methods (1), (2), (3) and (4) and negatively charged –COO– groups in the case of methods (6), (8) and (10). In the case of method (10), –COO– groups are additionally formed by the bonding of protein to the cyclic anhydride groups of the polymer. This polyvalent character of the protein carriers causes adsorption of the matrix, in a manner similar to ion exchangers, increased inactivation or denaturing of the protein structure and displacement of the pH optima in the case of bound enzymes.

There is, therefore, a great interest in the provision of a suitable carrier for water-insoluble proteins, as well as of water-insoluble proteins bound on a carrier and of a process for the preparation thereof.

This process substantially overcomes the deficiencies of prior carriers and/or techniques for binding proteins to carrier materials by providing copolymers eminently utilizable as protein carriers, methods for making same, a technique for binding proteins to such carriers, and the resulting protein/carrier combination as such.

The copolymers essentially consist of (a) acrylamide, together with (b) ethylene-maleic acid or its anhydride and/or (c) maleic acid, and also (d) N,N'-methylene-bisacrylamide or ethylene diacrylate in a weight ratio a:b:c:d of 3:0.5-1.5:0.05-4:0.075-0.9, the weight ratio a:(b + c) being not more than 4.

The residue of the dicarboxylic acid in the copolymer, i.e., that derived from component (b) and/or (c), is preferably present in the form of the anhydride. The weight ratio of acrylamide to component (d), i.e., N,N'-methylenebisacrylamide and/or ethylene diacrylate, is preferably 3 to not more than 0.45.

Acrylamide, N,N'-methylbisacrylamide and/or ethylene diacrylate, as well as maleic acid or ethylene-maleic acid or the corresponding anhydrides are polymerized in the presence of free-radical catalysts and free radical accelerators in the above given ratios and with the use of known methods. Although it is possible to carry out such a polymerization using maleic acid or maleic anhydride in wide ranges of concentration, copolymers suitable for binding proteins are only obtained when using the weight ratios given above.

In the production of the copolymer, the ratio of acrylamide to maleic acid or ethylene-maleic acid or to the corresponding anhydrides can be varied within the given ranges without substantially affecting the necessary amount of cross-linking agent (d). Thus, acrylamide can be polymerized with maleic acid or its anhydride in a ratio by weight of 3:01 to 4 without the weight ratio of acrylamide to cross-linking agent hereby being substantially influenced.

However, as indicated above, in the case of the use of ethylene-maleic acid or of its anhydride as comonomer, the limits are narrower. In this case, the content of cross-linking agent governs the pore size and strength, as well as the bonding of the ethylene-maleic acid in the gel.

For use for fixing of proteins, especially good copolymers have been obtained with the use of acrylamide and maleic acid or ethylene-maleic acid or the corresponding anhydrides in a ratio by weight of 3:05 to 1.0. In the case of increasing the concentration of the dicarboxylic acid to 1.5, the copolymer becomes more brittle and less swellable but still exhibits good properties as a carrier for proteins.

In the case of a ratio of acrylamide to cross-linking agent, i.e., N,N'-methylenebisacrylamide or ethylene diacrylate, of 3:0.075 to 0.450, there are obtained gel-like polymers with especially good properties. A ratio of 3:0.075 represents the lower limit in which the maleic acid or ethylene-maleic acid can still be satisfactorily incorporated into the copolymer.

For the fixing of proteins in aqueous media, there have also proved to be very useful copolymers with cross-linking agents in amounts corresponding to a weight ratio of acrylamide to cross-linking agent of more than 3:0.45, for example of 3:0.6 to 0.9. However, when using protein bound to such carriers at salt concentrations of more than 0.05 M, the relatively closely cross-linked gels hereby obtained have proved to be disadvantageous for the protein structure.

The production of the copolymers can be carried out, for example, in aqueous solution and preferably in an inert atmosphere. Examples of free radical catalysts or free radical accelerators which can be used include propionic acid nitrile and ammonium peroxydisulfate. The polymerization can be carried out at ambient temperature or also at higher temperatures of up to 100°C, as well as at reduced temperatures. The period of time of the polymerization depends upon the selected polymerization temperature, the selected accelerators or catalysts and on the composition by weight of the reaction mixture. Usually, it is between about 5 minutes and several hours.

After solidification, for the achievement of a particle size which is especially suitable for handling, the polymerized material is forced through a sieve with the desired mesh size, then washed with water and lyophilized. For the fixing of the protein, it is necessary that the dicarboxylic acid residues in the polymer be present in the form of the anhydride. Therefore, before the bonding of the protein, all acid groups present must again be converted into cyclic anhydride groups. This can be effected, for example, by heating the lyophilized copolymer in a vacuum to a temperature of about 80° to 120°C or at atmospheric pressure to a temperature of about 160° to 220°C, preferably of 180° to 200°C. The copolymers depending upon the concentration of cross-linking agent, are water white to milky cloudy substances with a rubber-like to brittle consistency.

When the carboxylic acid groups are converted into cyclic anhydride groups by heating to about 200°C, a slight yellow coloration frequently occurs. If heating is continued for a comparatively long period of time, then the color deepens from yellow to brown and, at the same time, the binding capacity decreases markedly. Therefore, heating for a comparatively long period of time at the necessary temperature under atmospheric pressure is to be avoided if possible.

The copolymer permits the binding of protein under mild conditions because it is hydrophilic and extremely swellable and the protein bonding only takes place by solvolysis of the cyclic anhydride groups. The protein structure is thus maintained after the bonding of the protein has taken place and activity yields of up to 100% and even more, due to a potentiation of the activity, can be obtained. Unbound protein can be recovered without loss of activity. In the case of the binding, the protein does not perform a cross-linking function in the matrix but rather the carrier-bound enzymes, which can be called "enzyme gels," are completely uniform and do not exhibit a differing strength dependent upon the concentration of the protein.

Due to its precise degree of cross-linking, the molecular sieve properties of the copolymer can be so adjusted that only proteins in the desired range can penetrate into it. Thus, it is possible so to cross-link the copolymer for certain proteins that the protein binding can only take place on the surface of the gel particles or, if desired, can also take place in the totality of the gel. The polymerization parameters can be varied according to the effective molecular radii, steric effects and charge distributions.

The copolymers possess adsorption properties such that, in water or weak buffer solutions (up to about 0.01 M), the protein is completely fixed on the carrier. Up to 85% of the protein is hereby covalently bound and the remainder heteropolarly bound.

The protein-binding capacity of the copolymer is dependent upon the number of anhydride groups present and upon the pore size. In the case of a more closely cross-linked product, the binding capacity is lower than in the case of a less closely cross-linked product with the same number of anhydride groups present since the protein molecule, in the case of closer cross-linking, cannot react all of the available groups capable of bonding.

If, in the copolymer too many charges are present which cannot be occupied by protein by covalent bonding, then generally there takes place a displacement of the pH optima in the case of the bound enzymes. Correspondingly, it has been observed, in the case of known carrier-bound enzymes, that, in the case of negatively-charged carried materials, a displacement of two pH units occured; for example, in the case of trypsin bound to ethylene-maleic acid anhydride, from 7.8 to 10.0 and, in the case of positively-charged carriers, a displacement into the acid region took place. In the case of the copolymers, it is possible to use the groups capable of bonding to a maximum extent so that carrier-bound proteins can be obtained which show no displacement of the pH value optimum.

Examples 1 and 2 illustrate the preparation of the copolymers, while Examples 3 and 4 give the preparation of the carrier-bound proteins.

Example 1: 3 g acrylamide and 0.3 g N,N'-methylenebisacrylamide were suspended, together with 1 g ethylene-maleic acid anhydride, in 23 ml 0.05 M phosphate buffer of pH 7.6 and, with vigorous stirring in an atmosphere of nitrogen and at ambient temperature, mixed with 1 ml of 5% propionic acid nitrile and 1 ml 5% ammonium peroxydisulfate.

After 10 to 15 minutes, the mass solidified to a gel-like block. This was left to stand for 60 minutes at ambient temperatures and then forced through a metal sieve with an internal mesh size of 0.5 mm. The granulate obtained was washed with about 5 liters of distilled water at this stage (the wet weight amounted to about 38.2 g).

The swollen copolymer thus obtained was lyophilized in order to deswell it. In a dry state, the weight of the copolymer amounted to 5.4 g. For cyclization to the acid anhydride, the material was subsequently heated for 1.5 to 2 hours at 200°C in a drying cabinet.

Example 2: 3 g acrylamide, 0.075 g ethylene diacrylate and 1 g maleic acid were copolymerized in the manner described in Example 1. The period of polymerization was 15 hours. Further working up was carried out in the manner described in Example 1. The yield of lyophilized product was 2.3 g. After cyclization, the weight of the product was 2.05 g.

Example 3: There was used a cyclized copolymer of 3 g acrylamide 0.15 g N,N'-methylenebisacrylamide and 1 g maleic acid as carrier. 100 mg trypsin were stirred with 1 g carrier in 15 ml ice-cold water (pH 5.2) and allowed to react for 15 hours at 4°C in a cool room. The product was subsequently filled into a column.

The heteropolar bound protein was eluted with 65 ml wash water. Thereafter, with the use of 0.2 M phosphate buffer (pH 7.8), no further protein could be eluted.

Heteropolar-bound protein:	31%
at pH 7.0, 8.5, 10:	20%
Residual activity, referred	
to nonbound trypsin:	50%

Example 4: The same carrier as in Example 3 was employed. 100 mg carrier and 200 mg trypsin were allowed to react in the manner described in Example 3. The nonbound protein was 50%; the activity of the bound protein before treatment with 0.2 M buffer was 100% and after treatment with 0.2 M buffer, 48%.

Acrylic Acid, Acrylamide, Methylenebisacrylamide Copolymer

M.A. Stahmann and Y. Ohno; U.S. Patent 3,536,587; October 27, 1970; assigned to Tejiin Limited, Japan have developed a process for the preparation of an enzyme resin by reacting an enzyme with an acid azide derivative derived from a hydrazide prepared from a copolymer consisting of 9 to 75 weight percent of a lower aliphatic ester of acrylic acid, 20 to 90 weight percent of acrylamide and 0.5 to 10 weight percent of N,N'-mono- or polymethylenebisacrylamide. The N,N'-mono- or polymethylenebisacrylamide is a compound expressed by the following formula

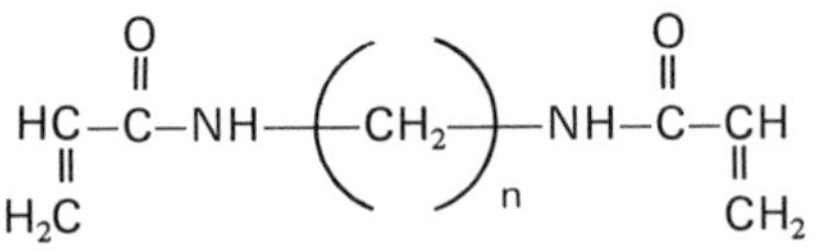

$$
\begin{array}{ccc}
 & O & & & O \\
 & \parallel & & & \parallel \\
HC-C-NH & -(CH_2)_n- & NH-C-CH \\
\parallel & & & & \parallel \\
H_2C & & & & CH_2
\end{array}
$$

wherein n is 1 to 30, or may be more preferably 1 to 6 (this compound will be called merely bisacrylamide hereinafter).

When bisacrylamide is used in an amount less than 0.5% by weight, the crosslinking is insufficient and part of the resulting enzyme resin becomes water-soluble. It may be used in an amount more than 10% by weight, but in this amount, it is difficult to control the particle size of the enzyme resin, therefore, the preferred amount of bisacrylamide is 1 to 5% by weight.

When the three components are copolymerized, copolymers consisting of acrylic acid units represented by the formula

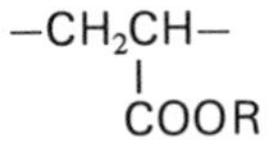

$$
\begin{array}{c}
-CH_2CH- \\
| \\
COOR
\end{array}
$$

acrylamide units represented by the formula

$$
\begin{array}{c}
-CH_2-CH- \\
| \\
CONH_2
\end{array}
$$

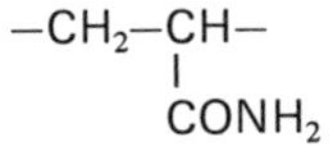

and methylenebisacrylamide units represented by the formula

$$
\begin{array}{ccc}
 & O & & & O \\
| & \parallel & & & \parallel & | \\
HC-C-NH & -(CH_2)_n- & NH-C-CH \\
| & & & & | \\
H_2C & & & & CH_2 \\
| & & & & |
\end{array}
$$

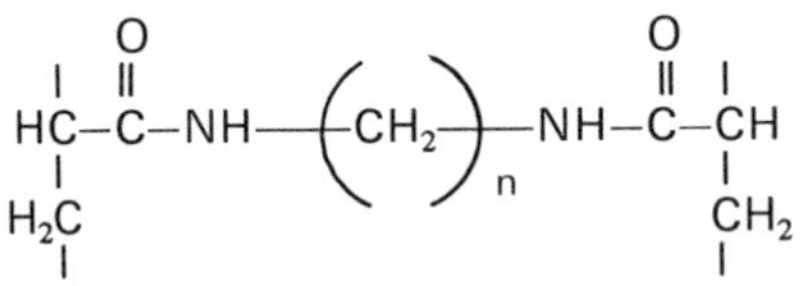

are obtained. In order to chemically combine any one of these copolymers with an enzyme, the acrylic acid units must be activated. As means for activation, the acrylic acid units are converted to azide by reacting copolymer with hydrazine and thereafter with nitrous acid. The enzyme resin can be also prepared by other processes. Instead of using the lower aliphatic esters of acrylic acid, the same amount of acrylic acid is used and after polymerizing the acrylic acid with the other two components mentioned above to obtain the copolymer. Thereafter the copolymer is activated by converting the acrylic acid units to an acrylic acid halide or an acrylic acid anhydride and then reacting the activated copolymer with an enzyme.

Because the enzyme resin is completely water-insoluble, it can be easily recovered by filtration or centrifugation after having been used in an enzymatic reaction, and reused. Furthermore, this enzyme resin is produced in spherical form which is suitable for packing into a column, and so the enzymatic reaction can be continuously carried out by the use of such enzyme resin.

If an enzyme resin column is used, a reaction solution flowing from it does not contain impurities derived from the enzyme, and it is not necessary to conduct a complicated post-treatment for removal of enzyme which is necessitated when the enzyme alone is used.

Acrylamide has an effect of stabilizing the enzyme, and a copolymer derivative comprising more than 20% by weight of acrylamide is especially effective, as shown in Table 1 below, which gives the acrylamide content in an acrylic copolymer and the enzymatic activity of an acylase resin obtained from it.

TABLE 1

Run No.	Acrylamide Content*	Activity Ratio**
1	0	0.6
2	10	0.6
3	20	0.8
4	48	1.0

Note — Substrate is N-acetyl-DL-alanine.
*Percent by weight
**Ratio of activity of the acylase resin with respect to the bound acylase and that of the starting acylase.

Example 1: Twelve parts of a hydrazide derivative derived from a copolymer consisting of 4% by weight of N,N'-methylenebisacrylamide, 48% by weight of acrylamide and 48% by weight of methyl acrylate was suspended in 600 parts by volume of distilled water, and 60 parts by volume of 36.5% hydrochloric acid and 300 parts by volume of a 4% aqueous solution of sodium nitrite were gradually added. This reaction mixture was stirred at 0°C for one hour.

Then, a water-insoluble azide derivative was quickly separated by filtration, washed several times with an ice-cooled water, and added to 1,200 parts by volume of a 0.1 M potassium phosphate buffer (pH 7.0) in which 1.2 parts of trypsin has been dissolved. The mixture was stirred continuously for 24 hours at 4°C. Unreacted trypsin was completely removed by washing five times with 1,000 parts by volume of a 0.02 M potassium phosphate buffer (pH 7.0).

There was 0.46 part of enzyme protein contained per 100 parts of the resulting enzyme resin. The hydrolytic activity of the enzyme resin measured by continuous titration of a liberated acid from N-α-benzoyl-L-arginine ethyl ester in an aqueous solution is 40% of that of the starting trypsin with respect to the amount of the coupled enzyme. Trypsin, when stored for one month at 4°C in an aqueous solution having a pH of 5, retained only 6% of the original enzymatic activity, while the enzyme resin retained 35% of the activity of the starting trypsin under the same conditions.

Example 2: Twelve parts of the water-insoluble azide derivative obtained under the same conditions as in Example 1 was reacted with 2.4 parts of acylase in 1,200 parts by volume of 0.5 M sodium bicarbonate solution (pH 8.2) to give 12 parts of a water-insoluble enzyme resin. There was 0.76 part of enzyme protein bound to 100 parts of this enzyme resin. When N-acetyl-α-amino acid was used as a substrate, it showed the following enzymatic activity with respect to the amount of bound enzyme, which gives the enzymatic activity of water-insoluble acylase resin.

TABLE 2

Substrate	Activity Ratio*
N-acetyl-DL-alanine	1.00
N-acetyl-DL-methionine	0.86
N-acetyl-DL-norleucine	0.79

*Ratio of activity of insoluble acylase resin to that of acylase.

Diamine Cross-Linked Sulfite-Polyacrolein Polymer

P.S. Forgione; U.S. Patent 3,753,861; August 21, 1973; assigned to American Cyanamid Company has demonstrated that enzymes can be covalently bound to carbonyl polymers, such as aldehyde and ketone polymers, in such a manner that the resultant covalently-bound enzyme remains catalytically active even after extensive and continuous use thereof in the conversion of enzymatically convertible substrates.

The compositions are hydrophilic, sulfited polymers having a catalytically active enzyme covalently bound thereto. In the case of most aldehyde and ketone polymers, the polymer is first treated with a suitable sulfite or material which imparts a sulfite group onto the polymer chain such as a sulfite per se, a hydrosulfite, a bisulfite, sulfurous acid, etc. The reaction is conducted at a temperature ranging from about 25° to about 90°C at atmospheric pressure.

After the sulfite treatment, the sulfited aldehyde or ketone polymers are then immobilized, e.g., insolubilized, if not already hydrophilic by chemically cross-linking them with a cross-linking agent or an immobilization agent. The sulfited polymer adduct or reaction product is contacted, for example, with a diamine such as ethylene diamine, tetramethylene diamine, N-methylethylene diamine, etc. at a temperature of about 0° to 150°C and in the presence of a solvent.

Insolubilizing or immobilizing the sulfited aldehyde or ketone polymer can also be accomplished in a multiplicity of other ways such as by vinyl cross-linking i.e., first producing an aldehyde or ketone polymer containing unsaturation and then reacting it, after sulfite treatment, with a polyunsaturated cross-linking agent such as divinyl benzene etc. Insolubilizing can also be effected by using any other polyfunctional compound which will cause the formation of a polymeric network via reaction with the sulfited polymer adduct through available sites such as vinyl groups, OH groups etc. Grafting of the sulfited polymer can also be accomplished to render the polymer immobilized.

Additionally, the insolubilization can be effected by reacting the sulfited polymer adduct with such agents as 4-aminophenyl sulfide hydrochloride salt etc. through available groups on the ketone or aldehyde polymers.

Polyacrolein, a water-insoluble polymer which contains some groups with which most enzymes are reactive, is first contacted with a sulfite such as sodium bisulfite in order to render it water-soluble and more susceptible to enzyme reaction. In such a condition, however, the polymer cannot be reacted with an enzyme because recovery of any product thereof is relatively impossible. Cross-linking of the sulfite-polymer product, however, renders it gel-like in consistency and effectively hydrophilic so as to allow reaction with the enzyme.

As a result, the sulfite-polyacrolein product is cross-linked with a diamine such as hexamethylene diamine. Based on theory, the result of these two reactions is that the sulfite breaks some of the heterocyclic rings of the polyacrolein creating more enzyme-reactive aldehyde groups thereon, in addition to a series of sulfite groups.

The diamine reacts with some of these aldehyde groups with the formation of —CH=N— linkages between two polymer molecules, thereby cross-linking the polyacrolein. Reaction of the enzyme, e.g., invertase, forms an adduct or covalent bond between the enzyme and the other available aldehyde groups, and also may result in reaction through the sulfite groups. The resultant adduct is then comprised of a series of cross-linked groups, free aldehyde groups, heterocyclic sulfite reaction groups and bound enzyme groups. One proposed structure of such a product is the following, sodium bisulfite and hexamethylene diamine having been used:

Example 1: Hydrophilic, Covalently Bound Invertase-Polyacrolein Adduct (Conversion of Sucrose to Invert Sugar) — The cross-linked polyacrolein-sodium bisulfite addition complex is prepared as follows. 0.88 part of 1,6-hexanediamine containing 3 parts of water, is slowly added, with stirring, to 44 parts of a 10% solution of polyacrolein-sodium bisulfite adduct (MW 80,000). The mixture is then heated to 65°C for 10 minutes and the yellow hydrophilic product which forms is then washed until neutral with distilled water and filtered.

The wet hydrophilic adduct thus obtained is suspended in 50 parts of water and reacted with 0.104 part of invertase (twice recrystallized) which had been first dissolved in 4 parts of water. The enzyme reaction mixture (pH 6.8) is gently stirred for 18 hours at 10°C and the resulting hydrophilic, covalently bound enzyme polymer adduct is then washed free of unreacted enzyme. Assay of the hydrophilic enzyme-polymer adduct shows high activity and about 80% of the charged enzyme has been bound.

Sucrose is converted to invert sugar as follows. 5% by weight of the isolated wet cake (72% H$_2$O) converts 100 ml of a 10% sucrose solution at pH 4.8 to 81.0% invert sugar in 15 minutes at 42°C. Similarly, 5% by weight of the freeze-dried polymer enzyme adduct prepared from the wet cake results in a 75.8% conversion to invert sugar using the same experimental conditions. Five months later the activity of the bound enzyme is still about 85% of its initial value.

Example 2: Hydrophilic, Covalently Bound Trypsin-Polyacrolein Adduct — 10 parts of wet (approximately 10% solids) cross-linked polyacrolein-sodium bisulfite addition complex (as prepared in Example 1) are suspended in 40 parts of water at pH 3.5 and reacted with 0.110 part of crystalline trypsin, dissolved in 4 parts of water.

The reaction mixture is stirred for 18 hours at 10°C and at the end of this time the resulting covalently bound trypsin adduct is washed until free of unbound enzyme. Analysis of the washings at 280 mμ indicates that 78.2% of the enzyme is bound to the polymer. The wet polymer enzyme cake is freeze-dried, resulting in a product having 0.087 part trypsin per part of dry solids.

The product, on assay, affords 14.4 units of activity per mg of enzyme (native enzyme 66 units/mg), therefore possessing 21.7% of the original activity. The analytical procedure used is the standard pH stat method (pH 7.9) using benzoyl arginine ethyl ester (BAEE) substrate. After 1 month at 4°C, a second analysis shows full retention of activity.

Example 3: Hydrophilic, Covalently Bound Trypsin-Polyacrolein Adduct — 10 parts of wet (approximately 10% solids) cross-linked polyacrolein-sodium bisulfite addition complex (as prepared in Example 1) are suspended in 50 parts of water at pH 9.0 and reacted with 0.100 part of crystalline trypsin, dissolved in 4 parts of water. The reaction mixture is stirred for 15 hours at 10°C, and at the end of this time the resulting covalently bound trypsin adduct is washed until free of unbound enzyme.

Analysis of the washings at 280 mμ indicates that 33% of the enzyme is bound to the polymer. The wet polymer enzyme cake is freeze dried, resulting in a product having 0.033 part of enzyme per part of solids. The product, on assay, affords 21.7 units of activity per mg of enzyme, or 32.4% of the original activity of native trypsin (native enzyme 66 units mg). The analytical procedure used is the pH stat method with BAEE substrate described in Example 2. Similar results are observed.

Example 4: Hydrophilic, Covalently Bound Glucose Oxidase-Poly(Isopropenyl-methyl Ketone) Adduct — 2.5 parts of poly(isopropenylmethyl ketone) are stirred with 25 parts of water containing 2.1 parts of potassium metabisulfite at pH 5.8 for 16 hours at 75°C over a nitrogen atmosphere. At the end of this time, 1.3 parts of 2(2-aminoethyl)-5(6)-aminoethylbicyclo-2.2.1-heptane in 5 parts of water are added to the above reaction mixture and stirred at 80°C for 25 minutes.

The hydrophilic adduct which results is washed to neutral pH and filtered. The wet cake is suspended in 30 parts of water at a pH of 6.1 and reacted with 0.025 part of glucose oxidase dissolved in 2 parts of water at 10°C for 18 hours. The resultant covalently bound enzyme is washed free of unbound glucose oxidase with cold distilled water and filtered.

Assay of 0.5% of the total wet cake (74% H_2O) using the standard pH stat procedure (pH 6.3) (substrate: 50 ml of 1% glucose solution with excess oxygen) in which the liberated gluconic acid is titrated with standard 0.01 N NaOH indicates that the activity of the adduct is 18.0 units/mg/min (native glucose oxidase assays 20.5 units/mg/min). Over a period of 1 month, continuous charging of a 10% glucose solution saturated with oxygen shows a 20% overall deactivation of the enzyme. 90% of the enzyme is bound.

Example 5: Hydrophilic, Covalently Bound Glucose Isomerase-Polyacrolein Adduct — 40 parts of wet (about 17% solids) cross-linked polyacrolein-sodium bisulfite addition complex (as prepared in Example 1) are suspended in 20 parts of 1 molar phosphate buffer, pH 7.1, containing 9,000 units of glucose isomerase and 5 parts of glucose. The reaction mixture is stirred for 18 hours at 45°C and at the end of this time the resulting covalently bound enzyme is washed until free of unbound enzyme.

Assay of 0.04% of the total wet cake (83% H_2O) using the standard colorimetric cysteine-sulfuric acid procedure for fructose, results in a product having a total of 1,600 units or 219 units activity per gram of dry support. After 4 months the bound enzyme loses only 50% of its activity when used in a continuous column process.

Example 6: Hydrophilic, Covalently Bound Glucoamylase-Polyacrolein Adduct — 10 parts of wet (20.1% solids) cross-linked polyacrolein-sodium bisulfite addition complex (as prepared in Example 1) are suspended in 30 parts of 0.1 M phosphate buffer pH 5.6 and reacted with 5 parts of a commercially available glucoamylase solution. The reaction mixture is stirred for 18 hours at 25°C and at the end of this time the resulting covalently bound enzyme is washed until free of unbound enzyme.

Assay of 0.05% of the total wet cake using the standard colorimetric procedure for reducing sugars indicates the activity of the product to be 890 units per gram of dry support. This activity represents 40 mg enzyme bound per gram of dry support (based on 22 units/mg native glucoamylase). This bound enzyme (80% H_2O) reacts with 100 ml of a 25% soluble solution with less than 20% overall deactivation after 30 days.

Example 7: Hydrophilic, Covalently Bound Catalase-Polyacrolein Adduct — 5 parts of wet (8.1% solids) cross-linked polyacrolein-sodium bisulfite addition complex (as prepared in Example 1) are suspended in 30 parts of water at pH 7.2 and contacted with 0.100 part of crystalline catalase. The reaction mixture is stirred for 18 hours at room temperature and at the end of this time the resulting co-valently bound catalase is washed with distilled water until free of unbound catalase.

Assay of 5% of the total wet cake (92% H_2O) using the standard iodometric procedure with hydrogen peroxide substrate indicates the activity of the adduct is 6,020 units/g dry support or 10.0 mg catalase per gram dry support (based on assay of native catalase which is 600 units/mg enzyme). Activity of the bound catalase drops only 13% after 1 month of continual use.

Dialdehyde Cross-Linked Carbonyl Polymers

P.S. Forgione; U.S. Patent 3,770,700; November 6, 1973; assigned to American Cyanamid Company has found that a greater quantity of enzyme can be covalently bound to carbonyl polymers, such as aldehyde and ketone polymers, if the carbonyl polymer is cross-linked, i.e., rendered insoluble, with a dialdehyde rather than other known cross-linking agents since the number of aldehyde groups available for enzyme binding is not thereby decreased.

Reacting the carbonyl polymer with the dialdehyde via an aldol condensation reaction, does not cause a decrease in the aldehyde groups, through which cross-linking occurs, available in the carbonyl polymer as results with the use of other cross-linkers because the aldehyde groups of the dialdehyde cross-linking agent either replace or add to those of the carbonyl polymer taken up in the cross-linking reaction or otherwise stabilize the total number of aldehyde groups present. As a result, the polymer becomes cross-linked and the number of available aldehyde groups on the cross-linked polymer remains substantially constant thereby enabling more enzyme to be bound.

The compositions comprise carbonyl polymers which are cross-linked with dialdehydes having the formula

$$\underset{H-C-R-C-H}{\overset{\displaystyle O \qquad\; O}{\overset{\displaystyle \|\;\;\;\;\;\; \|}{}}}$$

where R is a $+CH_2+_x$ group, x being 0 to 18, inclusive, a cyclic unsubstituted alkylene group of 4 to 10 carbon atoms, inclusive, or an unsubstituted arylene group of 6 to 12 carbon atoms, inclusive. The cross-linking reaction is an aldol condensation type of reaction and occurs between the reactive groups of the carbonyl polymer and the α-carbon atom of the dialdehyde.

The cross-linked polymer formed according to this process is insoluble and therefore precipitates from solution. Recovery of the polymer can be effected by filtration, centrifugation, etc. The cross-linked polymers are preferably prepared for the subsequent binding of catalytically active enzymes thereto. To this end, the carbonyl polymer must generally be first rendered susceptible to reaction with the enzyme which will be subsequently bound thereto. This is accomplished, in the case of most aldehyde polymers, by reaction of the polymer with a suitable solubilizing agent such as a bisulfite, specifically an alkali metal or alkaline earth metal bisulfite such as sodium, potassium, calcium, etc. bisulfite. The reaction is conducted at a temperature ranging from about 25° to about 90°C, at atmospheric pressure, although superatmospheric or subatmospheric pressure can be utilized.

After the bisulfite treatment, the aldehyde polymers are made hydrophilic by cross-linking as discussed above. By the term "hydrophilic" is meant that the polymer is wettable or swellable in water but is not substantially soluble. The compositions can contain hydrophobic members or portions provided that they also have hydrophilic portions which function as such when in contact with water.

If desired, the cross-linking step may be accomplished first and the bisulfite reaction second when the cross-linked polymer is to be used to bind an enzyme thereto, the only criteria being that the resultant product is in such a state so as to allow reaction with the enzyme.

When the aldehyde or ketone polymer is per se water-soluble, the bisulfite reaction need not be conducted and the enzyme can be contacted with the polymer immediately after rendering it insoluble, by cross-linking with the dialdehyde as discussed above. The basic requirement when enzyme binding is to be accomplished, is that the final enzyme-polymer adduct must be hydrophilic in order that it may be utilized in the enzymatic conversion of substrates to their conversion products.

Example 1: 4.0 parts of a 20% glutaraldehyde solution are slowly added with stirring to 40 parts of a 10% solution of polyacrolein-sodium bisulfite adduct (MW 80,000) previously adjusted to pH 10 with 10% sodium hydroxide solution. The mixture is stirred at 23°C for one hour and the off-white hydrophilic product which forms is then filtered and washed to neutrality.

The wet hydrophilic adduct thus obtained is suspended in 50 parts of water and reacted with 0.104 part of invertase (twice recrystallized) which has been first dissolved in 4 parts of water. The enzyme reaction mixture (pH 6.5) is gently stirred for 18 hours at 10°C and the resulting hydrophilic covalently bound enzyme polymer adduct is then washed free of unreacted enzyme. Assay of the hydrophilic enzyme-polymer adduct shows high activity with sucrose solution, in that 4%, by weight, of the isolated wet cake converts 100 ml of a 10% sucrose solution at pH 4.8 to 83.0% invert sugar in 15 min at 42°C.

Example 2: 5 parts of a copolymer of acrolein-styrene (93.8:6.2 weight ratio, respectively) are suspended in 40 parts of water containing 4.3 parts of sodium metabisulfite at pH 5.7. The mixture is stirred over nitrogen at 65°C for 5 hours, resulting in a completely water-soluble product.

To this bisulfite addition product is slowly added, with stirring, 4.0 parts of a 20% solution of adipaldehyde and the pH of the reaction mixture adjusted to pH 10.5 with 10% sodium hydroxide solution. The reaction mixture is stirred at 20°C for 2 hours. The resultant cross-linked, hydrophilic copolymer is filtered, washed to neutral pH, suspended in 20 parts of water and reacted with 4 ml of technical grade (k = 0.6) invertase at 18°C for 15 hours. The resulting covalently bound enzyme polymer adduct is washed free of unbound enzyme with distilled water and filtered (8.4 parts wet). Assay of 10%, by weight, of the wet product with 100 ml of 10% sucrose solution results in a 46.0% conversion to invert sugar.

Example 3: 10 parts of wet (approximately 10% solids) cross-linked polyacro-lein-sodium bisulfite addition complex (as prepared in Example 1) are suspended in 40 parts of water at pH 7.0 and reacted with 0.110 part of crystalline trypsin, dissolved in 4 parts of water. The reaction mixture is stirred for 18 hours at 10°C and at the end of this time the resulting covalently bound trypsin adduct is washed until free of unbound enzyme.

Analysis of the washings at 280 mμ indicates that 81.3% of the enzyme is bound to the polymer. The wet polymer enzyme cake is freeze dried, resulting in a product having 0.091 part trypsin per part of solids. The product, on assay, affords 18.4 units of activity per mg of enzyme (native enzyme 66 units/mg), therefore possessing 28.0% of the original activity. The analytical procedure used is the standard pH stat method (pH 7.9) using benzoyl arginine ethyl ester (BAEE) substrate.

Example 4: 2.5 parts of poly(isopropenylmethyl ketone) are stirred with 25 parts of water containing 2.1 parts of potassium metabisulfite at pH 5.8 for 16 hours at 75°C over a nitrogen atmosphere. At the end of this time, 1.5 parts of a 20% phthaladehyde solution are added to the above reaction mixture and stirred at 25°C for one hour. The hydrophilic adduct which results is washed to neutral pH and filtered. The wet cake is suspended in 30 parts of water and reacted with 0.025 part of glucose oxidase dissolved in 2 parts of water at 10°C for 18 hours.

The resultant covalently bound enzyme is washed free of unbound glucose oxi-dase with cold distilled water and filtered. Assay of 0.5% of the total wet cake using the standard pH stat procedure (pH 6.3) (substrate: 50 ml of 1% glucose solution with excess oxygen) in which the liberated gluconic acid is titrated with standard 0.01 N NaOH indicates that the activity of the adduct is 18.6 units/mg/min (native glucose oxidase assays 20.5 units/mg/min).

Protein-Diazotized Polyurethane

T.R. Sato; U.S. Patent 3,574,062; April 6, 1971; assigned to Beckman Instruments, Inc. has immobilized enzymes by reacting an enzyme with a diazo-protein substrate such as diazo-protein polyurethane. The substrate may be prepared by diazotizing a polyester polyurethane and coupling a protein to the diazotized polyurethane. The immobilized enzyme may be used as a catalyst in analytical procedures and the support is preferably a reticulated structure.

For use in an analytical procedure, it is preferred that the support permit a high flow of liquid such as aqueous solutions and for this reason polyurethane foams have been found particularly suitable as supports. Such immobilized enzymes have particular utility in biochemical analysis. When immobilized enzymes are used, there is a saving in the amount of enzyme required, a saving enhanced by the long periods during which such immobilized enzymes remain active; a substantial reduction in the size, weight and complexity of the analytical system since the enzyme solution reservoir, separation equipment and recycling equipment are eliminated, and since enzymes having higher specific activities can be made available, reaction times are shortened.

Example: A disk of polyester polyurethane foam having a thickness of one-eighth inch was washed with acetone to remove fatty substances. The disk was then washed with distilled water to remove residual acetone. Reticulation, i.e., removal of the membranes to leave a skeletal structure, was accomplished by treatment with 0.5 normal sodium hydroxide at 60°C for 5 to 10 min, followed by rinsing with distilled water.

The reticulated polyurethane was then subjected to a diazotization procedure. Four milliliters (3 mols) of sodium nitrite was added dropwise to an ice-cold solution of one gram of p-amino hippuric acid in 15 ml of 3 normal hydrochloric acid until the solution gave a positive test to starch-iodine paper. The temperature was maintained near 0°C by use of an ice-salt water bath. The diazonium salt solution was adjusted to pH 7.0 with sodium hydroxide.

The diazonium salt solution was then immediately added to the reticulated polyurethane disk which was squeezed to allow the solution to be imbibed by the polyurethane. After 12 hours the polyurethane diazo hippurated disk was thoroughly washed and stored under distilled water.

Crystallized bovine serum albumin was then coupled to the polyurethane diazo hippurate in the presence of N,N'-dicyclohexyl carbodiimide. The disk was added to a 25 ml solution at 0°C of borate buffer (pH 8.5) which contained 0.5 g of crystallized bovine serum albumin. A 10 ml solution (at 0°C) of tetrahydrofuran containing 0.5 g of N,N'-dicyclohexyl carbodiimide (DCC) was then added to the disk soaked with serum albumin. The disk was then squeezed several times in the DCC-containing solution to insure permeation of the disks with the albumin.

The disk was stirred in a beaker with the above mixture for 24 hours at 5°C and then pressed in a Buechner funnel to remove the solution after which it was exhaustively washed with a mixture of tetrahydrofuran and water.

A 10 ml solution (at 0°C) of borate buffer (pH 8.5) containing purified glucose oxidase (enzymatic activity of 25,000 units) and catalase (3,000 to 5,000 Katf units) was used to saturate the serum albumin-polyurethane diazo hippurate disk. A solution of DCC (1 g in 10 ml tetrahydrofuran) at about 0°C was added with squeezing of the disk to distribute the resultant solution throughout the disk.

The disk was then vigorously shaken in a vial with the enzyme mixture for 48 hours at 5°C. After shaking, the mixture was stirred for 16 hours to hydrolyze any unreacted DCC. After filtering in a Buechner funnel, the disk was washed extensively on the filter with a mixture of tetrahydrofuran and water to dissolve any unreacted DCC and some of the dicyclohexylurea which results from hydrolysis of the DCC. After washing, the disk was suspended in a dilute sodium bicarbonate solution and separated from the solvent by filtration. This was followed by washing with distilled water until glucose oxidase FAD is not detected.

The same procedure was used with other polyester polyurethane disks with the exception that p-amino phenyl acetic acid, p-amino-4-phenyl butyric acid and p-amino benzoic acid were each substituted for p-amino hippuric acid.

The products were tested for enzymatic activity and were compared with products produced by other methods. It was found that the product prepared using p-amino hippuric acid had an initial glucose oxidase activity of 1,120 units and activities of 920 and 760 units after 24 and 48 hours, respectively.

A product prepared in the same manner except that the use of serum albumin was omitted had an initial activity of 51 units and no activity after 24 and 48 hours while a product using albumin but omitting diazotization had an initial activity of 170 and no activity after 24 and 48 hours. A product prepared by reacting glucose oxidase with the polyurethane disk had an initial activity of 34 units and no activity after 24 and 48 hours.

Similarly, the product in which p-amino phenylacetic acid was used in diazotization had an initial activity of 905 units and activities of 751 and 624 units after 24 and 48 hours respectively while the same product prepared without use of a protein had an initial activity of 0.45 unit and no activity after 24 or 48 hours.

Freeze Drying of Immobilized Enzymes

P.S. Forgione; U.S. Patent 3,770,588; November 6, 1973; assigned to American Cyanamid Company has shown that the catalytic activity of enzymes bound to carriers can be maintained, stabilized or otherwise be prevented from deactivation by freeze drying the carrier bound enzyme from a solution of the substrate which is normally converted to its conversion product by the bound enzyme.

Carrier bound enzymes become deactivated upon storage, shipment etc. and in such a condition they cannot function in the normal (natural) manner so as to convert substrates to their conversion products. By freeze drying the carrier bound enzyme from a solution of the same substrate which is normally converted to its conversion product by the enzyme, the activity can be maintained even at temperatures which would normally deactivate the particular enzyme involved. At least about 1.0%, by weight, of substrate has generally been found to be sufficient, although higher concentrations, i.e., up to about 50%, by weight, can be used.

The carrier bound enzyme, being water-insoluble, is suspended in a substrate solution at a pH within 1.5 pH units of the optimum pH at which the substrate is catalytically acted upon by the enzyme. The pH may be adjusted to the appropriate value by any known means such as by adding acids such as dilute hydrochloric acid or bases such as dilute sodium hydroxide etc. and freeze dried.

Thus, for example, polymer bound invertase can be suspended in an aqueous sucrose solution at a pH of 4.8 ($\pm$ 1.5) and freeze dried to a powder consistency, stored and shipped and then be continually contacted therewith to convert the sucrose to invert sugar. The catalytic activity of the invertase has thereby been maintained for a longer period of time than if the material is not freeze dried.

Example 1: 5 Parts of a copolymer of acrolein-styrene (93.8:6.2 weight ratio, respectively) are suspended in 40 parts of water containing 4.3 parts of sodium metabisulfite at pH 5.7. The mixture is stirred over nitrogen at 65°C for 5 hours, resulting in a completely water-soluble product.

To this bisulfite addition product is slowly added, with stirring, 2 parts of 1,6-hexane diamine dissolved in 10 parts of water. The reaction mixture is stirred at 70°C for one hour. The resultant cross-linked, hydrophilic copolymer is filtered, washed to neutral pH, suspended in 20 parts of water and reacted with 4 ml of technical grade (k = 0.6) invertase at 18°C for 15 hours.

The resulting covalently bound enzyme polymer adduct is washed free of unbound enzyme with distilled water, filtered (7.97 parts wet), suspended in sucrose solution and freeze dried. The adduct maintains its catalytic activity at room temperature for 9 weeks.

Example 2: 10 Parts of wet (approximately 10% solids) cross-linked polyacrolein-sodium bisulfite addition complex are suspended in 40 parts of water at pH 3.5 and reacted with 0.110 part of crystalline trypsin, dissolved in 4 parts of water. The reaction mixture is stirred for 18 hours at 10°C and at the end of this time the resulting covalently bound trypsin adduct is washed until free of unbound enzyme.

Analysis of the washings at 280 mμ indicates that 78.2% of the enzyme is bound to the polymer. The wet polymer enzyme cake is suspended in a 5% aqueous casein solution at pH 8.1 and freeze dried. The product is recovered and, on assay, affords 17.2 units of activity per mg of enzyme (native enzyme 66 units per milligram), therefore possessing 26.1% of the original activity. The analytical procedure used is the standard pH stat method (pH 7.9) using benzoyl arginine ethyl ester (BAEE) substrate. Without freeze drying the polymer cake in casein solution but air drying, the activity ceases in 48 hours.

Example 3: 2.5 Parts of poly(isopropenylmethyl ketone) are stirred with 25 parts of water containing 2.1 parts of potassium metabisulfite at pH 5.8 for 16 hours at 75°C over a nitrogen atmosphere. At the end of this time, 1.3 parts of 2-(2-amino-ethyl)-5(6)-aminoethylbicyclo-2.2.1-heptane in 5 parts of water are added to the above reaction mixture and stirred at 80°C for 25 minutes.

The hydrophilic adduct which results is washed to neutral pH and filtered. The wet cake is suspended in 30 parts of water and reacted with 0.025 part of glucose oxidase dissolved in 2 parts of water at 10°C for 18 hours. The resultant covalently bound enzyme is washed free of unbound glucose oxidase with cold distilled water and filtered. A portion of the product is suspended in 10% aqueous glucose at pH 5.5 and freeze dried while a second portion is air dried.

Assay of 0.5% of the freeze dried powdery product 9 weeks later using the standard pH stat procedure (pH 6.3) (substrate: 50 ml of 1% glucose solution with excess oxygen) in which the liberated gluconic acid is titrated with standard 0.01 N NaOH indicates that the activity of the adduct is 17.7 units/mg/min (native glucose oxidase assays 20.5 units/mg/min). The air dried portion of originally the same activity is substantially depleted in 36 hours.

Example 4: 90 Parts of acrolein are reacted at room temperature with 10 parts of methylenebisacrylamide and 0.1 part of α,α'-azobis-α,α-dimethylvaleronitrile for 3 days under a nitrogen atmosphere. The resultant solid, cross-linked copolymer (8.3% weight percent bisacrylamide) is washed with water, filtered and dried at room temperature to constant weight.

10 Parts of the dried, cross-linked copolymer are then treated with 8.6 parts of sodium metabisulfite dissolved in 82 parts of water at pH 5.6 at 65°C with stirring for one day over nitrogen. The resultant hydrophilic bisulfite adduct is filtered and washed with water. The wet bisulfite adduct (147 parts) is suspended in 150 parts of water and treated with 12 ml of technical invertase solution (k = 0.6) with stirring at 10°C for 18 hours.

The reaction mixture is washed free of unbound invertase and assayed for activity. 5% of the wet product converts 100 ml of 10% sucrose solution to 32% invert sugar. A conversion of 30% is recorded utilizing a freeze dried powder produced from a 10% aqueous sucrose solution thereof. An air dried portion of the product shows no activity after 30 hours at ambient temperature.

Following the techniques of the previous examples, various other enzymes are covalently bound to polymeric materials to produce a hydrophilic enzyme composition which is then freeze dried from its suitable substrate solution. The results are set forth below.

Enzyme bound*	Substrate solution from which freeze-dried	Results 6 weeks after freeze-drying
glucose isomerase.......	20% aqueous invert sugar solution — pH 7.5.	Converts glucose to fructose.
glucoamylase	15% solution of liquified starch in water — pH 4.2.	Converts polysaccharides to glucose.
α-amylase	10% solution of liquified starch in water — pH 6.9.	Continuously hydrolyzes polysaccharides to glucose and maltose.
tyrosinase	2% tyrosine in water-glycerol (90/10) solution — pH 6.5.	Tyrosine converted continuously to dihydroxyphenylalanine.
pectinase...................	1% pectin in water-methanol (98/2) solution — pH 8.0.	Acts on pectin continuously to form pectic acid.

*In each instance, air dried polymer bound enzyme lost its activity after 24 hr.

Immobilization in an Epoxy-Containing Polymer

J. S. Matthews; U.S. Patent 3,821,084; June 28, 1974; assigned to Gulf Research & Development Company has reported a method for synthetically immobilizing and supporting enzymes on an epoxy containing polymer. In this method an enzyme or mixture of enzymes is chemically united with an epoxy containing polymer in the presence of an amine. Not only does the enzyme which is immobilized by the epoxy containing polymer in the presence of an amine retain its natural activity to a substantial degree, but it can also be repeatedly reused with substantial retention of its activity.

The immobilized enzyme is chemically bound or coupled to the epoxy containing polymer through one or more epoxy groups of the support or carrier. This coupling reaction takes place with a suitable functional group in the enzyme molecule. Since an epoxy group can react with many different functional groups, the epoxy containing polymer carrier is reactive with and can immobilize a large number and variety of enzymes for catalytic utility.

The epoxy containing polymer by which the enzyme is immobilized is prepared by the reaction of an epoxy monomer having at least one 1,2-epoxy group and one terminal unsaturation per molecule when the epoxy monomer is a single chemical species or per average molecule when the epoxy monomer is a mixture of chemical species as a result of its method of preparation. Suitable available epoxy monomers include glycidyl acrylate, glycidyl methacrylate, allyl glycidyl ether, and the like.

Other epoxy monomers which are useful herein include those which are prepared by the reaction of an epoxy resin with a difunctional olefin to form the epoxy monomer as defined. The epoxy monomer is then copolymerized with an olefin monomer by free radical polymerization to form the epoxy containing polymer. In the final step this epoxy containing polymer and the desired enzyme are reacted in the presence of the amine to immobilize the enzyme on the polymer support.

The amine can suitably be a primary, secondary or tertiary amine having the formula $N(R_5)(R_6)(R_7)$ in which R_5 is alkyl having from one to six carbon atoms, phenyl or benzyl and R_6 and R_7 are optionally hydrogen or a hydrocarbon group as defined by R_5 or a cyclic amine such as piperidine or pyridine. Examples of suitable amines include trimethylamine, triethylamine, tributylamine, mono-ethylamine, mono-n-hexylamine, aniline, benzylamine, diethylamine, and the like.

The expressions epoxy resin and polyepoxide are used interchangeably to refer to the broad class of initial epoxide containing reactants which are used in the reaction of the difunctional olefin with the epoxy resin to form the epoxy monomer. The epoxy resin can be a single compound or a mixture of compounds containing the alpha-epoxy group and can be monomeric or polymeric. Each epoxy group can be located terminally, internally, or on a cyclic structure that is capable of being used in preparing a thermoset material.

The expressions are used with reference to the thermoplastic or uncured state and do not refer to a thermoset or cured material. When the epoxy resin is a single compound, it must contain at least two epoxy groups per molecule. However, with epoxy resins or polyepoxides in which a variety of molecular species are present, the number of epoxy groups will vary from molecular species to molecular species such that the average number of epoxy groups per molecule is specified.

This average number of epoxy groups per molecule is also designated the epoxy equivalent value. When a mixture of compounds is involved, the epoxy equivalent value must be greater than one and preferably at least about two but will generally not be a whole integer. The epoxy equivalent value is obtained by dividing the average molecular weight of the epoxy resin by its epoxide equivalent weight (grams of epoxy resin containing one gram equivalent of epoxide). The epoxy resin can be aliphatic, cycloaliphatic, aromatic, heterocyclic, mixtures of these, saturated or unsaturated, and can include noninterfering groups.

The difunctional olefin can be defined by the following structural formula
$CH_2=C(R_1)R_2$ where R_1 is hydrogen or methyl and R_2 is carboxyl; hydroxymethyl;
formyl; chlorocarbonyl; carbamyl; aminomethyl; mercaptocarbonyl; mercapto-
methyl; $-CH_2NHR_3$ where R_3 is lower alkyl, hydroxyphenyl or lower alkyl substituted hydroxyphenyl, or lower alkyl or phenyl substituted hydroxyphenyl.

Lower alkyl refers to alkyl groups having one to four carbon atoms. Suitable
difunctional olefins include acrylic acid, methacrylic acid, allyl alcohol, acrolein,
methacrolein, acrylyl chloride, acrylamide, allylamine, thioacrylic acid, allyl mercaptan, vinyl phenol, and the like.

The difunctional olefin reacts through the functional group other than the unsaturated group with an epoxy group in the polyepoxide in order to form the
epoxy monomer. This reaction is illustrated with acrylic acid and a diepoxy resin:

$$CH_2{=}CH{-}\overset{\overset{O}{\|}}{C}{-}OH \;+\; CH_2{-}\overset{O}{\overset{/\backslash}{CH}}{-}[R]{-}CH{-}\overset{O}{\overset{/\backslash}{CH_2}} \longrightarrow CH_2{=}CH{-}\overset{\overset{O}{\|}}{C}{-}O{-}CH_2{-}\overset{\overset{OH}{|}}{CH}{-}[R]{-}CH{-}\overset{O}{\overset{/\backslash}{CH_2}}$$

where R merely represents the nonepoxy portion of the polyepoxide. Since the
preparation of an epoxy containing polymer is the objective, it is preferable that
the amount of the difunctional olefin relative to the epoxy resin be less than the
amount that would react with every epoxy group in order that unreacted epoxy
groups are available for subsequent reaction after the polymer is formed.

Therefore, the ratio of mol equivalents of epoxy groups to difunctional olefin
should preferably be greater than one. It is preferred that this ratio be at least
about two to one in the reaction mixture used to make the epoxy monomer.
This is simply accomplished by using an equimolar mixture of epoxy resin and
difunctional olefin in which the epoxy resin has an epoxy equivalent value of at
least about two.

A catalyst is preferably used for the desired reaction between the epoxy resin
and the difunctional olefin. Suitable catalysts for this reaction are the alkyl and
aromatic tertiary amines including trimethylamine, triethylamine, tripropylamine,
tributylamine, benzyldimethylamine, benzyldiethylamine, pyridine, 2-picoline,
4-picoline, 2,6-lutidine, and the like. The reflux is carried out until the reaction
is substantially complete. The epoxy monomer is next copolymerized with the
olefinic monomer.

The olefinic monomer copolymerizes with the epoxy monomer by free radical
polymerization at the olefinic double bonds in each material. The olefinic monomers are suitable for copolymerization with the epoxy monomer to produce hydrophilic epoxy containing polymers without substantial reaction with the epoxy
groups in the epoxy monomer.

The preferred olefinic monomers can be defined by the structural formula
$CH_2=C(R_1)R_4$ where R_1 is methyl or hydrogen and R_4 is cyano, lower carboalkoxy
having one to four carbon alkoxy, and the like and include acrylonitrile, methacrylonitrile, methyl acrylate, methyl methacrylate, and the like. Since the olefinic monomer introduces the hydrophilic property into the polymer, the relative

amount of this monomer can be varied to adjust this property. Suitable hydrophilic epoxy containing polymers are made when the epoxy monomer comprises about 5 to about 30 mol percent of the comonomer mixture.

Suitable free radical initiation can be used such as ionizing radiation, ultraviolet radiation and the like, but preferably chemical free radical initiators are used. The chemical free radical initiators together with accelerators or activators, if needed, are selected according to common practice by correlating the desired temperature of polymerization with the activation temperature of the initiators.

Suitable chemical free radical initiators include benzoyl peroxide, lauroyl peroxide, methyl ethyl ketone peroxide, di(2-methylpentanoyl)peroxide, p-chlorobenzoyl peroxide, cyclohexanone peroxide, bis(1-hydroxycyclohexyl)peroxide, hydroxyheptyl peroxide, and the like, dicyclohexyl peroxydicarbonate, dibenzyl peroxydicarbonate, azobisisobutyronitrile, and the like. The polymerization reaction is carried out using a suitable, nonreactive solvent for the monomers generally at a temperature between about 80° and about 200°C. Suitable solvents include the solvents specified for the preparation of the epoxy monomer. The same solvent can be conveniently used in each reaction. Following polymerization the polymer is dried of the solvent and formed into a fine powder.

The polymer product is a solid hydrophilic polymer which contains a substantial number of epoxy groups at the surface of the particles available for reaction with other compositions. In order to bind an enzyme to the water-insoluble epoxy containing polymer, the polymer is first dispersed in a water solution of an enzyme and then the amine is added. If the amine is added to the enzyme before the polymer, the enzyme is substantially deactivated yet when the opposite order of addition is employed, the activity of the immobilized enzyme is substantially enhanced.

The amine is desirably used in an amount from about 10 to about 100 times the weight of the enzyme and preferably from about 20 to about 50 times the weight of the enzyme. Generally the weight of the enzyme in solution is no greater than about 1% of the weight of the polymer support.

In the upper end of this range, the enzyme is present in excess over that amount which can be bound. However, the enzyme does not need to be used in substantial excess since only a sufficient amount of enzyme need be used in the immobilization reaction to produce a product of useful activity.

During the immobilization reaction, the enzyme comes into reactive contact with the polymer aided by the hydrophilic nature of the polymer. The binding reaction occurs through one or more of the large number of epoxy groups available for reaction on the surface of the polymer particles and through one or more epoxy reactive groups in the enzyme.

Examples 1 through 8: Two epoxy containing polymers were made as illustrated by the following procedure. A 27 gram portion of Epon 834 (a diglycidyl ether of bisphenol A having an epoxy equivalent value of 280 obtained from Shell Chemical Company) dissolved in 100 ml of toluene, 8.2 grams of methacrylic acid and 1.0 ml of triethylamine were placed in the resin kettle equipped with mechanical stirrer, condenser, nitrogen inlet and heating mantle. The reaction mixture was refluxed with stirring under a nitrogen atmosphere as one ml samples of the reaction mixture were periodically sampled and titrated with 0.1 N sodium hydroxide to analyze for methacrylic acid. When analysis indicated that about 95% of the methacrylic acid had been consumed, 50 ml of toluene were removed by distillation.

This was followed by the addition of 200 ml of benzene and 27 g of methacrylonitrile and one gram of azobisisobutyronitrile. The reaction mixture was stirred under reflux for 5 hours during which time the polymer product came out of solution. The polymer product was filtered, washed with benzene, ground and dried under vacuum yielding 61% based on total reactants (38 grams) of a fluffy, white powdered polymer product having an epoxy equivalent value of 3,817.

A 500 mg portion of polymer which had passed through a 100 mesh sieve was placed in a 30 ml screw cap bottle. Glucose oxidase, which had been dissolved in a 0.1 M potassium phosphate buffer at pH 7.0, was added to the bottle. The amine was added to the bottle which was closed and slowly rotated in a cold room at about $0°C$ for 4 hours. The suspension was then filtered through a medium porosity fritted glass funnel and washed with about 90 ml of phosphate buffer to remove unbound enzyme. The filtrate was diluted and colorimetrically analyzed for unbound enzyme. The bound enzyme was determined by difference. An aliquot of the damp polymer was analyzed and the activity determined colorimetrically.

The data set out in the following table compares the activity of 6 immobilized glucose oxidase products in which an amine was used in the immobilization reaction and two where no amine was used. Polymer A was made by the procedure described above in which 20 mol percent of the epoxy monomer was polymerized with methacrylonitrile while polymer B differed in that it was made by polymerizing ten mol percent of the epoxy monomer with methacrylonitrile. In each example 500 mg of the polymer was used.

Example	Polymer	Amine	E/A*	Bound Enzyme mg/g**	Activity, %***
1	A	none	0.58/0	0.66	26
2	A	n-tributyl	0.58/16	0.52	43
3	A	piperidine	0.58/16	0.54	33
4	A	pyridine	0.58/16	0.52	31
5	B	none	1.94/0	1.96	16
6	B	triethyl	1.94/7	2.10	26
7	B	triethyl	0.98/56	1.54	53
8	B	triethyl	0.98/280	0.98	24

*mg of glucose oxidase and mg of amine used in the immobilization solution.

**mg of immobilized glucose oxidase per g of polymer.

***activity of immobilized enzyme x 100/activity of enzyme in solution.

When 56 mg of triethylamine were added to the enzyme solution described in Example 7 without polymer present, the activity of the enzyme was reduced from 80.6 I.U. to 3.6 I.U. which is 4.5% of the original activity in solution.

INORGANIC CARRIERS AS SUPPORTS

Application of an Intermediate Silane Coupling Agent

An enzyme is generally considered a biological catalyst capable of initiating, promoting, and governing a chemical reaction without being used up in the process or becoming part of the product formed. It is a substance synthesized by plants, animals, some viruses and microorganisms. All enzymes isolated thus far have been found to be proteins, i.e., peptide polymers of amino acids. An enzyme may contain prosthetic groups such as flavin adenine dinucleotide, porphyrin, diphosphopyridine nucleotide, etc. Most enzymes are macromolecules, generally, having a molecular weight greater than 6,000.

The specificity of enzymes and their ability to catalyze reactions of substrates at low concentrations have been of particular interest in chemical analyses. Enzyme catalyzed reactions have been used for some time for the qualitative and quantitative determination of substrates, activators, inhibitors, and also enzymes themselves.

Until recently, the disadvantages arising from the use of enzymes have seriously limited their usefulness. Objections to the use of enzymes have been their instability, since they are susceptible to all the conditions which normally denature proteins, e.g., high temperature, concentration dependence, pH changes, microbial attack, and autohydrolysis. Furthermore, the cost of large amounts of enzymes has made their use in routine chemical analyses impractical.

Attempts have been made to prepare enzymes in an immobilized form without loss of activity so that one sample could be used continuously for many hours. The immobilized enzymes perform with increased accuracy all the operations as those of ordinary soluble enzymes: that is, they can be used to determine the concentration of a substrate, of an enzyme inhibitor, or of an enzyme activator.

These have been made by physically entrapping enzymes in starch gel, polyacrylamide gel, agar, etc. Enzymes have been insolubilized by diazotizing them to cellulose derivatives and to polyaminostyrene beads. Enzymes have also been insolubilized on polytyrosyl polypeptides and collodion matrices. The main disadvantages of using such organic materials are (a) that they are subject to microbial attack resulting from the presence of carbon atoms in the polymer chain whereby the carrier is broken down and the enzymes solubilized; (b) substrate diffusion in many cases becomes the limiting factor in reaction velocity thereby decreasing apparent enzyme activity; and (c) when employed in chromatographic columns, the pH and solvent conditions increase or decrease swelling affecting flow rates of the substrate through the column.

R.A. Messing and H.H. Weetall; U.S. Patent 3,519,538; July 7, 1970; assigned to Corning Glass Works have described a method of stabilizing enzymes by chemically coupling the enzymes to inorganic carriers which are substantially immune to attack by microbial organisms. The enzyme is chemically coupled to the carrier by a silane coupling agent which by proper selection substantially reduces or eliminates entirely loss of activity due to interference with the active sites of the enzyme molecule. Highly stable enzymes are prepared by this technique which can be used and reused over extended periods of time. It is thus even possible to calibrate the activity of an enzyme and have some certainty that upon reuse, its activity level will be substantially constant. These stabilized enzymes find considerable use in analytic procedures and may also be used in the preparation of chemicals, pharmaceuticals, and foodstuffs.

The carriers are inorganic materials having available oxide or hydroxide groups. These materials must be substantially water insoluble and are either weak acids or weak bases. They may also be classified in terms of chemical composition as siliceous materials or nonsiliceous metal oxides.

Of the siliceous materials, a preferred carrier is porous glass either in particulate form or as an integral piece such as a disc. Glass has the advantage in that it is dimensionally stable and that it can be thoroughly cleaned to remove contaminants as for example by sterilization. Other siliceous inorganic carriers which can also be used include colloidal silica, wollastonite (a natural occurring calcium silicate), dried silica gel, and bentonite. Representative nonsiliceous metal oxides include alumina, hydroxy apatite, and nickel oxide.

The silane coupling agents are molecules which possess two different kinds of reactivity. These are organofunctional and silicon-functional silicon compounds characterized in that the silicon portion of the molecule has an affinity for inorganic materials such as glass and aluminum silicate, while the organic portion of the molecule is tailored to combine with many organics.

The main function of the coupling agent is to provide a bond between the enzyme (organic) and the carrier (inorganic). In theory, the variety of possible organofunctional silanes is limited only by the number of known organofunctional groups and the available sites on the enzyme molecule for bonding. A multitude of different silane coupling agents can be used as illustrated by the general formula:

$$(Y'R')_n Si R_{4-n}$$

where Y' is a member selected from the group consisting of amino, carbonyl, carboxy, isocyano, diazo, isothiocyano, nitroso, sulfhydryl, halocarbonyl; R is a member selected from the group consisting of lower alkoxy, phenoxy, and halo; R' is a member selected from the group consisting of lower alkyl, lower alkyl-phenyl, and phenyl; and n is an integer having a value of 1 to 3. Useful silane coupling agents may be represented by the formula

$$Y_n Si R_{4-n}$$

where Y is a member selected from the group consisting of amino, carbonyl, carboxyl, hydroxyphenyl, and sulfhydryl; R is a member selected from the group consisting of lower alkoxy, phenoxy, and halo; and n is an integer having a value of 1 to 3. However, most available coupling agents have the following formula.

$$RCH_2CH_2-Si(OCH_3)_3$$

Where R is a reactive organic group, tailored to match the reactivity of the system in which it is to be used.

Important types of bonding between the coupling agent and the enzyme illustrated merely by their functional or reactive groups, are as follows:

Types of Bonding

Bond Type	Bond structure	Reactive groups	
		Enzyme	Coupling agent
(1) Amide	$-\overset{O}{\underset{\|}{C}}-NH-$	$-COOH$ $-NH_2$	$-NH_2$ $-COOH$
(2) Sulfonamides	$-N-\overset{S}{\underset{\|}{C}}-NH-$	$-NH_4$	$-NH_2+ClCCl$ (with $\overset{S}{\|}$)
(3) Azo linkage	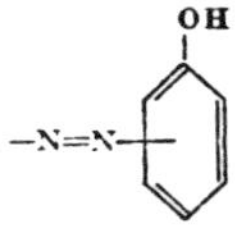 $-N=N-\langle\text{phenyl, OH}\rangle$	Tyrosine Histidine Lysine	$-N_2^+Cl^-$
(4) Ether	$R-O-R$	$-C-ONa$	$-R-X$
(5) Ester	$-\overset{O}{\underset{\|}{C}}-O-R$	$-COOH$	$-R-OH$
(6) Disulfide	$R-S-S-R$	$R-SH$	$R-SH$

Useful coupling agents are amino-functional aliphatic silanes such as N-beta-amino-ethyl-gamma-aminopropyl trimethoxysilane, N-beta-aminoethyl-(alpha-methyl-gamma-aminopropyl)-dimethoxymethylsilane, and gamma-aminopropyl-triethoxy-silane.

In order to select the optimum coupling agent or agents, it is important to consider the active sites on the enzyme molecule. Thus, it is undesirable to bond an enzyme having an amine group in its active site by means of an amine-silicate bond. Consequently, a coupling agent should be selected which is nondestructive to the enzyme, as for example in trypsin, by bonding to the carboxyl or sulfhydryl group of the enzyme.

Furthermore, the coupling agent must be such that bonding can be produced under conditions (e.g., temperature and pH) that they do not destroy either the enzyme or the carrier. The conditions under which the bonded enzyme is to be used are also significant in that the type of bond formed between the coupling agent and the enzyme, which to a large extent depends on the selection of coupling agent, should be stable at those conditions.

The bonding of the enzyme to the carrier is principally a two step reaction. Briefly, the first step involves bonding the coupling agent to the carrier and the second step involves bonding the enzyme to the coupling agent-carrier combination. The quantity of enzyme coupled appears to be dependent upon the surface area of the carrier available for reaction. Enzyme activity is dependent upon mildness of coupling conditions, but not necessarily the structure of the active site.

Example 1: A sample of powdered porous 96% silica glass (950 ± 50 A pore size, 16 m^2/g surface area) was washed in 0.2 N HNO$_3$ at 80°C with continuous sonication for at least 3 hours. The glass was washed several times with distilled water by decantation and then heated to 625°C overnight in the presence of O$_2$.

The glass was cooled and placed into a round bottomed flask. To each 2 g of glass were added 100 ml of a 10% solution of γ-aminopropyltriethoxysilane in toluene. The mixture was refluxed overnight and washed with acetone. The final product was air dried and stored. The resultant aminoalkylsilane derivative was found to contain 0.171 meq of silane residues/g of glass as determined by total nitrogen.

One gram of treated glass was added to 3.5 ml of distilled water containing 100 mg of crystalline trypsin. This was then added to a mixture of 0.5 ml N,N'-dicyclohexylcarbodiimide (DCCI) in 0.5 ml tetrahydrofuran (THF). The reactants were stirred overnight at room temperature. The product was washed exhaustively with NaHCO$_3$ solution, 0.001 M HCl, and distilled water. The insolubilized trypsin was stored in 0.001 M HCl at 5°C.

The hydrolysis of benzoyl-arginine ethyl ester (BAEE) was carried out at pH 8.1 in 0.1 M glycine. One unit of activity is equal to the hydrolysis of 1 μmol of substrate/minute at 25°C at pH 8.1.

Three representative samples assayed by the above method were found to contain 8.5 units, 25.2 units, and 12.0 units per gram of glass. This represents only microgram quantities of active enzyme. The total enzyme coupled by this method was 0.347 mg/g glass as determined by total nitrogen.

Example 2: To 2 g of aminoalkylsilane derivative of porous glass (780 ± 50 A pore size) was added 1 g of p-nitrobenzoic acid. This was stirred for two days at room temperature in a 10% solution of DCCI in absolute methanol. The reacted material was washed exhaustively in methanol, added to 500 ml of distilled water containing 5.0 g sodium dithionite and boiled for 30 minutes. The p-aminobenzoic acid amide of the aminoalkylsilane-glass (aminoarylsilane derivative) was washed with distilled water, followed by acetone and air dried.

The aminoarylsilane derivative was diazotized in 0.1 N HCl by addition of an excess of solid NaNO$_2$ at 0°C. One gram of product (diazoarylsilane derivative) was added to 14 mg of crystalline trypsin in 50 ml NaHCO$_3$ solution. The reaction was continued at 5°C overnight. Thereafter, the chemically coupled trypsin was washed in NaHCO$_3$ solution and distilled water.

The assay was carried out as described previously except that the substrate contained 0.08 mg BAEE/ml dissolved in 0.07 M phosphate buffer adjusted to pH 7.0. The chemically coupled trypsin contained the equivalent of 0.189 mg of active enzyme/g of glass. Nitrogen determination revealed 5.67 mg enzyme coupled per g glass. The enzyme activity retained was 3.2% of the total trypsin coupled.

To show the stability of the chemically coupled trypsin as compared to the free (uncoupled) trypsin, a comparative experiment was run at 23°C. Crystalline trypsin at a concentration of 0.5 mg/ml was placed in 0.07 M phosphate buffer solution. At intervals 0.05 samples were withdrawn, added to 3.0 ml of the substrate, and assayed for activity.

Within two hours, all enzyme activity was destroyed by autohydrolysis of the free trypsin. The original conversion rate of the chemically coupled enzyme was arbitarily set at 100% activity. No loss in activity was observed for 154 hours of continuous assay. Thereafter the enzyme began to lose activity, but even after 397 hours considerable activity was still observed.

Examples 3 thorugh 15: Various enzymes were chemically coupled to a series of representative inorganic carriers. In the table below the starting materials, namely the enzyme, the carrier, and the coupling agent used are indicated in columns 2, 3, and 4. The coupling agents are given in terms of letters wherein B is the diazoarylsilane derivative and C is the isothiocyanoalkylsilane derivative. The activities of the chemically coupled enzymes are shown in column 6 using the substrate of column 5.

Example	Coupled Enzyme	Inorganic Carrier	Coupling Agent	Substrate	Sample Activities, mg/g
3	Papain	Porous glass	C	Casein	1.5
4	Alkaline protease	Colloidal silica	C	Casein	92.0
5	Ficin	Porous glass	B	Casein	0.9
6	Ficin	Porous glass	C	Casein	2.7
7	Urease	Porous glass	B	Urea	1.0
8	Alkaline phosphate	Porous glass	B	p-Nitrophenyl-phosphate	0.7
9	Glucose oxidase	Porous glass	B	Dextrose	10.1
10	Glucose oxidase	Porous glass	C	Dextrose	11.8
11	Glucose oxidase	Colloidal silica	C	Dextrose	24.0
12	Glucose oxidase	Alumina	C	Dextrose	6.0
13	Glucose oxidase	Hydroxy apatite	C	Dextrose	10.7
14	Peroxidase	Porous glass	B	Hydrogen peroxide	1.0
15	Peroxidase	Nickel oxide	C	Hydrogen peroxide	0.5

Carrier Containing Reactive Silanol Groups

R.A. Messing; U.S. Patent 3,556,945; January 19, 1971; assigned to Corning Glass Works has stabilized enzymes by coupling the enzymes to an inorganic carrier containing reactive silanol groups through hydrogen and amine-silicate bonding whereby the enzyme becomes insoluble and can be used and reused over an extended period of time. The enzyme is strongly bonded to the carrier along the length of the chain.

The bonding of the enzyme to the carrier is a single step reaction that does not employ a coupling agent. This reaction involves hydrogen bonding and ionic bonding by means of the amine group of the enzyme and the silanol group of the ceramic carrier. More specifically, hydrogen bonding may occur through functional groups on the enzyme, e.g., carbonyl, amide, sulfhydryl and hydroxyl groups, while ionic type bonding is an amine-silicate bond at terminal or residual amines of the enzyme.

Enzyme powder is dissolved in a buffer solution and assayed. The enzyme is then bonded to a pretreated surface reactive ceramic carrier generally at below room temperature. The excess enzyme is removed and the bonded enzyme is rinsed with distilled water and buffer. Thereafter the bonded enzyme-carrier is leached and assayed until a stabilized value is obtained and finally the stabilized enzyme is stored in water or buffer at room temperature or below.

In bonding the enzyme to the ceramic carrier the aqueous enzyme solution is placed in contact with the carrier at a temperature below room temperature preferably about 5°C. After remaining in contact with the carrier for about 1 to 72 hours, the stabilized enzyme is bound to the ceramic surface and any excess enzyme is removed.

The initial uptake of protein by glass is isoelectric point dependent. The higher the isoelectric point of the molecule, that is, the more basic the protein because of the presence of amine groups, the greater the amount of protein bound in the initial reaction which is the chemical bonding of basic amine groups to reactive silanol groups of the glass. It is important that the pH of the solution during bonding is such that the enzyme does not denature, preferably this should be on the acid side of the isoelectric point.

It is important to leach out unbound enzyme before a stabilized assay value is obtained. Leaching may be a matter of hours and even up to many days. The length of the leaching period appears to be a function of the pore size of the ceramic. The small pore diameter ceramics, that is pores approaching the molecular dimensions of the enzyme, require longer leaching periods of from about 6 to 16 days.

The large pore material, such as fritted glass and wollastonite demonstrate little or no leaching periods. This appears to indicate that the leaching is a result of the combination of loosely bound enzyme within the pore and reduced diffusion rates due to the approach of the pore dimensions to the molecular dimensions of the enzyme. An induced negative charge across the first layer of bound enzyme may loosely bond second and third enzyme layers.

In preparing the carrier for bonding of the enzyme, it is frequently necessary to pretreat the carrier for the purpose of making the silanol groups more readily available and reactive. If, for example, the porous glass is not pretreated, it may contain many different contaminating substances which occupy the available bonding sites. By cleaning the surface of the glass, substantially all the active sites become available.

Example 1: A representative nuclease enzyme designated as RNAase having the following properties was prepared: molecular weight 12,700, isoelectric point pH 7.8, activity of preparation 1530 units/mg.

Assay conditions: 4 minute incubation at 37°C in 0.1 M acetate buffer pH 5.0. Substrate — 1% ribonucleic acid in buffer. Precipitant — 0.75% uranyl acetate in 25% perchloric acid.

Unit activity definition: one unit is equivalent to the amount of acid soluble oligonucleotide which causes an increase of spectrophotometric absorption at 260 mμ of 1.0 in four minutes at pH 5.0.

The carrier was a porous glass sample of Code 7930, cylinder, OD 1.2 cm, ID 1.0 cm, length 2 inches, weight 2.0 grams, pore diameter 74 A, surface area 113 m^2/g. Sample was cleaned by soaking in 1.2 M HCl for 30 minutes. The sample was then transferred to a furnace and the temperature was raised slowly to 550°C and maintained at that temperature for 3 hours. After cooling, the glass was immersed in and equilibrated against 0.1 M acetate buffer, pH 5.0 overnight.

Bonding of enzyme: The glass cylinder was placed in a test tube. The tube was immersed in a 5°C water bath, and 6 ml of solution containing 0.5 mg RNAase/ml in 0.1 M acetate buffer, pH 5.0 was added to the tube. The enzyme was permitted to react with the cylinder for 6 hours at 5°C. The cylinder was removed from the enzyme solution and extracted twice with the buffer.

Assay and storage RNAase cylinder: The cylinder was assayed in 20 ml of a 1:1 acetate buffer diluted ribonucleic acid solution. After 4 minutes of incubation a 2 ml aliquot was removed for analysis. Between analyses the cylinder was first washed thoroughly with the buffer and then stored in the buffer at 5°C.

The results of this study indicated an initial activity of 0.040 mg/g of glass. After a leaching period of approximately 12 days, the cylinder reaches a relatively constant activity level equivalent to 0.0075 mg RNAase per gram of glass. The activity remained at this level over a 111-day period.

The protein bound to the glass was calculated by determining optical densities at 280 mμ of the original enzyme solution, the solution used for binding and the extracts. Initially, 5% of the bound enzyme was active. This figure fell to 1.2% after the leaching period.

Example 2: A representative proteolytic enzyme, papain, was prepared having the following properties: molecular weight 21,000, isoelectric point pH 8.75, activity of preparation 0.01 units/mg.

Assay conditions (modified Kunitz procedure): 60 minute incubation at 40°C in 0.1 M phosphate buffer, pH 5.8, which contained 0.005 M cysteine and 0.001 M ethylenediamine tetraacetic acid (EDTA). Substrate — 1% boiled casein in buffer. Precipitant — 5% trichloroacetic acid.

Unit activity definition: one unit is equivalent to the amount of acid soluble peptide producing a change in optical density at 280 mμ of 0.001 per minute.

The carrier was a porous glass sample of Code 7930, cylinder, OD 1.2 cm, ID 1.0 cm, length 1 inch, weight 1.27 grams, pore diameter 68 A, surface area 118 m^2/g. Sample was ultrasonically cleaned in 0.2 N HNO_3 at 80°C. The sample was transferred to an oven and the temperature was slowly raised to 100°C and maintained at this temperature for 1 hour. The glass was then transferred to a water dessicator to equilibrate against water vapor. The glass was not buffer equilibrated before use.

Bonding of enzyme: The glass cylinder was placed in a test tube; the tube was immersed in a 5°C bath, and 4 ml of 0.1 M phosphate buffer, pH 6.95 containing 4 mg papain 1 ml was added. The enzyme was allowed to react with the cylinder for 1½ hours at 5°C. The cylinder was removed from the enzyme solution and extracted twice with water.

Assay and storage of papain cylinder: The cylinder was assayed in 5 ml of a 1:1 buffer diluted substrate solution containing 0.005 M cysteine and 0.001 M ethylenediamine tetraacetic acid. After 1 hour of incubation a 4 ml aliquot was withdrawn for analysis. Between analyses the cylinder was extracted with water and then either stored in water or 0.1 M phosphate buffer, pH 5.8 containing 0.01 M cysteine and 0.002 M ethylenediamine tetraacetic acid at 5°C (the solution used for storage had no apparent effect on the results).

After a leaching period of approximately 16 days, the cylinder reaches a relatively constant activity level equivalent to 0.07 mg papain per gram of glass. The activity remained at this level over a 122-day period. Initially 90% of the bound enzyme was active. This figure fell to 4% after the leaching period.

Example 3: Using glucose oxidase, the following experiment was performed. Porous glass sample: particles − 60 to +80 mesh, weight 1.0 gram, pore diameter 900 A, surface area 20 m^2/g. The sample was cleaned by exposing it to 625°C in the presence of oxygen for 15 minutes. The sample was cooled in a stream of O_2 and then equilibrated under reduced pressure against water vapor. The glass was not buffer equilibrated before use.

Bonding of enzyme: The glucose oxidase was bonded to the glass from 0.1 M phosphate buffer, pH 6.95. One gram of glass particles was exposed to 4 ml of enzyme solution (5 mg glucose oxidase/ml) for 70 hours at 5°C. The glass was extracted two times with buffer and then with water. The enzyme-glass particles were transferred to a piece of filter paper and the paper was bound with thread. The paper wrapped particles were used as a "tea-bag" for enzyme determinations.

Assay and storage of paper wrapped glucose oxidase particles: The tea-bag was assayed in 125 ml of 0.01 M phosphate buffer, pH 6.95, solution containing 18 mg glucose/ml at room temperature. 2.5 ml aliquots were removed at 3 min intervals for analysis. Between analyses the tea-bag was stored at 5°C in 0.01 M phosphate buffer, pH 6.95.

Over the initial 17 days of storage no apparent leaching period was noted and the glass appeared to exhibit a constant level of activity equivalent to 0.123 mg glucose oxidase per gram of glass.

Bonding of Enzymes in Presence of Substrates

R.A. Messing; U.S. Patent 3,802,997; April 9, 1974; assigned to Corning Glass Works has described a method of bonding enzymes to inorganic carriers which eliminates or substantially reduces the loss of enzyme activity. This method involves bonding the enzyme to the carrier in the presence of its substrate and thus apparently blocking the active sites of the enzyme to avoid reaction of these sites with the carrier. Highly stable enzymes can be prepared. These stabilized enzymes find considerable use in analytical chemistry and may also be used in the preparation of chemicals, pharmaceuticals, foodstuffs and the detergent industry.

The carriers are inorganic materials having available oxide and/or hydroxide groups. These materials must be substantially water insoluble and are either weak acids or weak bases. They may also be classified in terms of chemical composition as siliceous materials or nonsiliceous metal oxides. Of the siliceous materials, a preferred carrier is porous glass either in particulate form or as an integral piece such as a disc. Glass has the advantage in that it is dimensionally stable and that it can be thoroughly cleaned to remove contaminants as, for example, by sterilization.

Other siliceous inorganic carriers which can also be used include colloidal silica (CAB-O-SIL), wollastonite (a natural occurring calcium silicate), dried silica gel, and bentonite. Representative nonsiliceous metal oxides include alumina and hydroxy apatite. These representative inorganic carriers may be classified as shown in the table below.

Siliceous		Non-Siliceous Metal Oxides	
Amorphous	Crystalline	Acid MeO	Base MeO
Glass	Bentonite	Al_2O_3	Hydroxy
Silica Gel	Wollastonite		Apatite
Colloidal			
Silica			

In order to form the highly insolubilized enzyme and to prevent loss of enzyme activity, a substrate of the particular enzyme must be present during the bonding procedure. The importance of this may be illustrated in the bonding of trypsin wherein by forming an enzyme-substrate complex prior to the carrier, the amino groups in the active sites, i.e., nitrogen in the histidine residue, is occupied by the substrate.

Masking these groups prevents reaction between the active sites of the enzyme and the carrier and leaves the sites available for future enzymatic reactions. The substrate additionally functions as a cushion to prevent deformation of the enzyme molecule and as a complexing agent to reinforce the internal bonds of the enzyme. It is essential that the substrate be tailored to the specific enzyme.

The amount of substrate present during bonding should be sufficient to protect the enzyme during bonding. Usually an equal amount by weight of enzyme to substrate may be used as a rough examination. Substrates can be used alone or in combination as long as they are compatible. But when two or more different types of enzymes are to be bonded, best results will be obtained when a substrate for each type of enzyme is present.

The accompanying flow sheet, Figure 1.1, essentially illustrates the process. A full discussion of the drawing is set forth hereinbelow.

The carrier may be as formed bodies or as particulate materials. The carrier is

FIGURE 1.1: BONDING OF ENZYMES IN PRESENCE OF SUBSTRATES

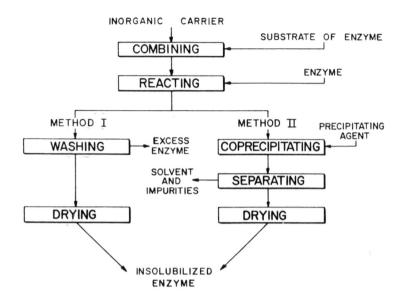

Source: R.A. Messing; U.S. Patent 3,802,997; April 9, 1974

first combined with the substrate usually in an aqueous suspension or solution. Temperature is not critical, but room temperature or slightly above is recommended. The time depends on the substrate and the temperature and generally low molecular weight substrates and higher temperatures result in shorter combining times, whereas higher molecular weight substrates and lower temperatures result in longer combining times. The term "combining" is broadly used to include saturating, coating, reacting, complexing, mixing, and dispersing. After the carrier and substrate have been combined, excess substrate may be removed.

In the next step, the enzyme in the presence of its substrate is reacted with the carrier. Actually, the reaction may be considered in two parts: initially the enzyme reacts with the substrate and thereafter the available residues of the enzyme not reacting with the substrate are permitted to react with the carrier.

Generally, the enzyme in aqueous solution is added to the carrier-substrate combination. The temperature of the reaction should be very low preferably about 5°C or below. The pH at which the reaction is conducted is also important and it is recommended that the pH be either higher or lower than the optimum pH of the enzyme substrate reaction. The reaction should be allowed to continue for a sufficient time, typically at least 20 minutes, to bond the enzyme to the carrier. The product, bonded enzyme in an aqueous medium, may be stored and used as desired.

However, for most purposes the excess enzyme is removed by filtration or centrifugation and washing in distilled water. Finally, the bonded enzyme is dried by conventional techniques such as by drying in air or vacuum, by spray drying, or freeze drying. The dried bonded enzyme may also be stored for use.

As an alternative procedure for stabilizing the enzyme, the carrier is in the form of fine particulate material preferably having a maximum particle size of 150 microns. The carrier is initially combined and mixed with the substrate in an aqueous suspension whereby the carrier (having a surface charge) is dispersed in the suspension.

The conditions of temperature and time are similar to the first method. The enzyme is then reacted with the carrier-substrate combination for a very short time, usually less than 5 minutes, and at low temperatures. Again there is an initial reaction between the enzyme and its substrate which is followed by the reaction between the enzyme and the carrier.

It is at this point that the enzyme and the carrier are coprecipitated by the addition of a precipitating agent. Coprecipitation may be either by dehydration or charge neutralization. In dehydration, an organic solvent such as acetone or an alcohol acts as the precipitating agent. Charge neutralization results from the addition of a salt solution, e.g., ammonium sulfate and sodium sulfate, to neutralize the charge on the protein molecules and the carrier particles. The temperature during coprecipitation should be generally below room temperature.

Finally, the precipitating agent is removed by such conventional procedures as filtration, centrifuging, washing, and air drying. The product obtained is a dried stabilized enzyme composite which is initially water insoluble.

Example 1: A quantity of powdered porous 96% silica glass (595 A pore size, 80 to 140 mesh U.S. Standard Sieve) was washed in 0.2 N HNO_3 with continuous sonication for at least 30 minutes. The glass was washed several times with distilled water and heat cycled to 625°C under an O_2 stream for 30 minutes. Thereafter, the glass was cooled in the O_2 atmosphere. A 0.2% gelatin substrate solution was prepared by adding 100 ml distilled water to 200 mg of gelatin (USP granular, 270 Bloom) and dissolving with heat and sonication.

To 500 mg of porous glass was added 5 ml of the gelatin solution and the mixture agitated at 37°C for 1½ hours. The residual gelatin solution (3.6 ml) was separated from the glass by decantation.

The gelatin protected glass sample was cooled to 5°C in a water bath and then 5 ml of an aqueous solution containing 1.39 mg/ml of alkaline protease from *B. subtilis* (Alcalase) was added. The enzyme was allowed to react with the glass at 5°C for 17½ hours. The enzyme solution was removed by filtration and the glass particles were thoroughly washed with water and air dried.

A control sample was also prepared in which 500 mg of porous glass was added to 1.4 ml of distilled water to approximate the liquid volume remaining in the gelatin protected sample after decantation. Thereafter, the bonding procedure was similar to that of the sample in which the enzyme was bound in the presence

of its substrate. The enzyme solutions and washes for each sample were analyzed for unbound protein at a wavelength of 280 mμ. The weights of the reacted glass and the amounts of enzyme bound (by difference at 280 mμ) were
as follows.

Sample	Dried Sample	Enzyme Protein Bound	Enzyme mg/g
Substrate Protected	1.023 g.	2.36 mg	2.31
Control	1.020 g.	2.78 mg	2.73

The bound enzyme was then assayed as follows:

Anson Hemoglobin*

Sample	mg E activity/g Sample	Active Enzyme
Substrate Protected	1.32	72%
Control	0.92	30%

*pH 9.72, 10 min, 37°C.

Azocoll*

Sample	mg E. Activity/g Sample	Active Enzyme
Substrate Protected	1.32	57%
Control	0.92	34%

*pH 10, 15 min, 37°C.

These results clearly show that the substrate protected system has substantially
greater activity than the control and, therefore, the substrate protected system
reduces the loss of enzymatic activity during bonding of the enzyme to an inorganic carrier.

Example 2: Following the procedure of Example 1, 200 mg of the alkaline protease enzyme was bonded to 600 mg of amorphous colloidal silica particles
(CAB-O-SIL) in the presence and absence of 200 mg of the protective gelatin
substrate. The bound samples were assayed with a hemoglobin substrate at pH
9.7 with 10 minutes and 45 minutes incubation at 55°C. The increased recovery
of the bound enzyme relative to the free enzyme at the longer incubation time
(45 minutes) is due to the gradual release of enzyme into solution with time

while the enzyme in free solution is destroyed by denaturation. The results were as follows:

Sample	Incubation Time	mg E Activity/g Sample	Active Enzyme
Substrate Protected	10 minutes	38.8	34.0%
	45 minutes	73.3	64.3%
Control	10 minutes	4.8	1.1%
	45 minutes	9.2	2.2%

Example 3: Following the procedure of Example 1, 0.6 gram of bentonite which had been heated to 500°C in O_2 for 1½ hours was added to 10 ml of a 2% gelatin solution. After the carrier and substrate were thoroughly mixed, 10 ml of a 2% (200 mg) alkaline protease solution was added at room temperature. The mixture of carrier, substrate, and enzyme was cooled to 5°C and allowed to react overnight. The product was filtered on a Buchner funnel, washed with distilled water, and air dried.

The enzyme protein bound to the bentonite was determined from the residual protein (spectrophotometrically at 280 mμ) in the wash and the filtrate. A control sample was prepared by adding 10 ml of the 2% (200 mg) alkaline protease solution to 1.0 g of bentonite and allowing the mixture to react overnight. The control was then treated in the same manner as the substrate protected sample. The weight of the reacted bentonite and the amount of enzyme bound were as follows:

Sample	Weight	mg Enzyme Protein Bound	mg Enzyme/g Sample
Substrate Protected	0.6870 g	86	125
Control	1.1244 g	146	130

The bound enzyme was then assayed by Anson Hemoglobin (pH 9.75, 10 minutes, 37°C)

Sample	mg E Activity/g Sample	Active Enzyme
Substrate Protected	40.0	32%
Control	1.2	0.9%

Porous Glass Carrier with a Metal Oxide Surface Layer

Various methods have been developed to insolubilize or immobilize normally soluble enzymes in such manners that the enzymes retain their activity, can be readily removed from a reaction solution, and used repeatedly. Enzymes have been insolubilized and immobilized in a variety of ways. For example, enzyme composites have been made by physically entrapping enzymes in such organic materials as starch gel, polyacrylamide gel, agar and the like. They have also been insolubilized by chemically coupling them via azo linkage to cellulose derivatives and to polyaminostyrene beads. Enzymes have also been insolubilized on polytyrosyl polypeptides and in collodion matrices.

Several disadvantages have been associated with using the organic materials. The organic materials have been found subject to microbial attack resulting from the presence of carbon atoms in the polymer chain whereby the carrier is broken down and the enzymes become solubilized. Also, many organic carriers have poor heat stability and thus, are difficult to sterilize by such means as an autoclave.

Further, some organic materials lack dimensional stability when used in columns and an increase or decrease in swelling of the materials can affect the flow rates of a substrate flowing through the column or the conformation of an attached enzyme. The above disadvantages, and others, associated with using organic carriers or organic matrices for insolubilizing or immobilizing enzymes led to the development of enzyme composites utilizing inorganic carriers.

W.H. Tomb and H.H. Weetall; U.S. Patent 3,783,101; January 1, 1974; assigned to Corning Glass Works have developed methods of preparing immobilized enzyme composites having a calculated half-life of up to 645 days. The significant increase in half-life is thought to be attributable to an improved inorganic carrier which is more durable in an aqueous environment than past inorganic carriers. The new carrier is prepared by coating a less water-durable base carrier with a more water-durable metal oxide which comprises a continuous metal oxide layer on the surface of the base carrier.

The preferred coating treatment comprises the steps of coating the base carrier with a solution containing a metal, the oxide of which is more water-durable than the base carrier, and then firing the coated carrier at between 200° and 800°C for about one to four hours, depending on the metal oxide to be formed, thereby forming a continuous metal oxide surface layer on the base carrier. Base inorganic carrier refers to any organic carrier which is less water-durable than a metal oxide(s) which can be used to coat it.

When porous glass is used as a base carrier, a preferred final coat consists of a thin, continuous surface layer of zirconium oxide. Such a surface metal oxide layer can be formed by first coating the base carrier with a solution containing zirconium in either an organic or aqueous solution, and then firing the coated base carrier to leave a thin, continuous surface layer of zirconium oxide. This surface layer may be silanized and modified for chemical coupling to enzymes by known means.

Preferably, the base carriers are porous materials (e.g., porous glass) so that a large surface area per unit weight and volume of carrier is available. The base carriers are thinly coated with a treating agent which enhances the durability of the carrier in an aqueous environment. Having increased durability, the coated carriers will have less solubility in water. This lower solubility is thought to be responsible for the increase in half-life of enzyme composites utilizing the coated inorganic carriers. With less dissolution of the carrier, there is less release of the coupled enzyme into a solution. With less release of the enzyme, there is an overall increase in the half-life of the composite since the enzyme is less likely to deactivate when maintained in an immobilized or insolubilized state.

The improved carriers are prepared by forming a thin, continuous layer of at least one metal oxide on the surface of the base carrier so that the durability of the coated carrier in an aqueous environment is better than the durability of the base carrier. By limiting the more durable coating to the surface of the base carrier, small amounts of relatively expensive coating agents can be used to coat a relatively less expensive base carrier which comprises the bulk of the overall weight of the treated carrier.

For example, when the base carrier is porous glass, any metal oxide coating (or a mixture of metal oxide coatings) which will form a thin, continuous layer on the surface can be used as long as it provides an overall increase in durability. Also, a durable metal oxide coating on an extremely porous base carrier having a high surface area yields a product having both high surface area and greater durability.

Thus, in those cases where a durable metal oxide cannot be prepared in a form having a high surface area, a coating of the same oxide on a base material having such a surface area provides a carrier having both desired properties. Among the metal oxides which can be used to coat a base carrier such as porous glass (mainly amorphous silica) are the oxides of such metals as zirconium, titanium, aluminum, hafnium, and like metals, the oxides of which are more water-durable than porous glass.

It is thought that metal oxide coatings or combinations of coatings for base carriers other than porous glass can be readily determined and prepared by one skilled in the art since the solubilities of various base carrier materials can be found in appropriate handbooks. Once the solubility of the base carrier is known, a more water-durable metal oxide, or mixture of oxides, can be chosen as the coating material. Such coating material can be applied by known methods such as by applying or contacting a solution of the metal to the base carrier and then firing the treated carrier to leave the oxide of the metal. Alternatively, the coatings may be applied by the illustrative detailed methods described below.

To apply the coating oxides in a thin, continuous layer, the metal may be applied as a metallo-organic compound such as, for example, a salt in an organic solution, an inorganic ester, or a chelate of the metal, which, after application, is convertible by further treatment to the metal oxide. The conversion of the metallo-organic compound to the respective metal oxide may be accomplished by heating the coated carrier to burn off the organic material, thus leaving a metal oxide coating which will enhance the durability of the base carrier.

Zirconium oxide may also be applied to the base carrier by using an aqueous solution. For example, zirconyl chloride or nitrate may be used as a starting material to coat the base carrier by dissolving about 30 g $ZrOCl_2$ or zirconyl nitrate in 250 ml H_2O for each 100 g of porous glass. This solution is mixed with the porous base carrier which is then placed under a vacuum for about 15 minutes to remove air from the pores and allow the solution to enter the pores. The treated porous carrier is then dried overnight at an elevated temperature (e.g., 145°C) and then fired at about 375°C for three hours. This treatment results in a coated carrier having a ZrO$_2$ coating comprising about 8% by weight of the coated carrier.

A coated carrier comprises a porous, granular, inorganic material which has a thin, continuous, surface layer of zirconia or titania. To prepare such carriers, a porous inorganic base material (e.g., porous glass particles or beads, 40 to 80 mesh, 500 A average pore diameter) is first dried to free it of surface water that is not chemically combined in the base carrier.

A solution of a chelated metal such as zirconium or titanium in an organic solvent is prepared (typically containing the equivalent of 8% oxide). This solution may be prepared by diluting Tyzor AA (Du Pont) with methanol if a coating consisting essentially of a titania oxide is desired, or the solution may be made by dissolving in methanol equimolar quantities of zirconium acetonate and tetra-n-propyl zirconate if a coating consisting essentially of a zirconium oxide is desired. The dry base carrier is then saturated with the solution at room temperature and then carefully dried at or above the boiling point of the organic solvent. Preferably, contact with moisture is avoided and the sample is dried with dry air.

After drying, the impregnated material is placed in a furnace and heated to a temperature of about 200° to 800°C, depending on the coating solution and the base carrier, and held thereat for about one to four hours. Preferably, the temperature is raised by about only 50°C per hour. While still warm, the material is removed from the furnace and stored in sealed jars until ready for silanization and enzyme coupling. At this point, the treated carrier consists of the base carrier having a thin, continuous coating of essentially the metal oxide or oxides chosen to coat the base carrier.

The oxide groups, being on the surface, are available for bonding to the silicon portion of a silane coupling agent. A silane coupling agent may be coupled to treated carrier with the result that the silanized treated carrier can be generally represented as

Base carrier-MeO-Silane

where Me represents the metal, the oxide of which forms the thin continuous layer over the base carrier.

The organo-functional portion of the silane coupling agent may be appropriately modified for the chemical coupling of the enzyme which is to be insolubilized or immobilized. For example, the organo-functional groups of the silane may be modified for diazotization and then diazotized for azo coupling to the enzyme if an azo-type coupling will not generally interfere with the enzyme's active sites.

Also, glutaraldehyde may be used to modify the silane for enzyme coupling. As pointed out in the above patent, the preferred silane-to-enzyme bond will depend on the active sites sought to be protected in a particular enzyme. Once the enzyme is coupled via the intermediate silane coupling agent to the coated base carrier, it can be represented by the formula shown on the following page

Base carrier-MeO-Silane-Enzyme

where Me represents the metal of the metal oxide coating of the base carrier, or a mixture of metal oxides, having an overall greater durability than the base carrier. The silane portion of the silane coupling agent is attached to the coated carrier, the metal oxide layer being disposed between the base carrier and the silane coupling agent. The organic portion of the coupling agent is coupled to the enzyme.

Detailed directions for preparing some preferred composites are given below. The enzymes glucoamylase and glucose isomerase were chemically coupled to the treated carriers by means of azo linkage or glutaraldehyde reaction in essentially the same manner as for the control composites. Preparation of the composites preferably involves a six-step procedure, as shown in the illustrative steps A through F below.

(A) Drying of CPG: 20 gram portions (dry basis) of CPG glass (40 to 80 mesh, pore volume 1.5 ml per gram) are placed in a covered 350 ml evaporating dish and removed to a circulating air oven operating at 200°C. After two hours, the samples are removed and placed in a desiccator. The desiccator is pumped to a pressure of 1½ inches of Hg or less and this pressure maintained until the sample ceases "boiling". The vacuum is then broken with dry nitrogen. The sample is then returned to the drying oven for another two hours or more and then again placed in the desiccator and the pressure reduced by means of a pump to 1½ inches of Hg or less.

If boiling occurs a second time, pumping is continued until it ceases, the vacuum broken with dry nitrogen, and the sample is again returned to the drying oven for two hours or more. If no boiling occurs, the desiccator is washed once with air, and the vacuum broken with dry nitrogen and the sample is allowed to cool to room temperature. The sample is weighed when it is removed from the desiccator and it is then ready for impregnation with the coating solutions. The drying steps above are merely illustrative and other techniques may be used depending on the material dried and equipment available.

(B) Preparation of Coating Solutions: As noted above, a wide variety of metal oxides may be used as a coating for the base carrier since the only requirement for the metal oxide(s) surface layer is that it is more durable in an aqueous environment than the base carrier.

Thus, various forms of a given metal or combination of metals such as salts, chelates, inorganic esters, and the like may be used in either an aqueous or organic solution to coat the base carrier. This coating may then be converted to a surface metal oxide layer by heat treatment or other techniques which will leave a metal oxide surface. It is thought that other methods of applying a metal oxide surface layer to an inorganic body could be used to prepare the coated carriers.

For example, it is well known that metals can be applied to surfaces by vapor deposition, sputtering, or similar techniques. Such metal deposits can then be converted to metal oxides by a variety of means such as, for example, by heat treatment. In the specific examples below, however, the coating was applied as a metal chelate solution which, after firing, left a metal oxide surface layer of zirconium oxide. The detailed methods for preparing zirconium chelate coating solutions are merely illustrative. Two alternative methods for preparing such a solution are given. The coating solution of the first method was the one used in the subsequent steps which resulted in the enzyme composites having coated carriers.

(1) 8% chelate solution #40 — 15.8 grams of zirconium acetylacetonate are dissolved in 69 grams of methanol. To this are added 15.2 grams of 75% tetra-n-propyl zirconate in n-propanol. While considerable heat evolves, the solution stays below boiling and is used immediately because of the tendency for a precipitate to separate slowly. The precipitate is believed to be tetramethyl zirconate.

(2) 8% chelate solution #66 — 13 grams of 2.4 pentanedione are dissolved in 56.6 grams of benzene and to this with stirring are added 30.4 grams of 75% tetra-n-propyl zirconate. While boiling should not result, the heat of reaction is higher than for #40 chelate solution and the reaction should be carried out in an unstoppered flask. This solution is stable and after cooling can be stored in a stoppered bottle for a week at least.

(C) Impregnation of the Dried CPG: Approximately 65 grams of the 8% chelate (of method 1, above) was weighed into a small beaker, and added from the beaker with stirring to a 20 g portion of dried CPG. The sample was saturated without flooding.

The sample was covered and dried for about at least six hours in a circulating oven at 80°C. Violent bumping may occur if the oven temperature is about 90°C or higher.

(D) Firing of the Impregnated and Dried CPG: The impregnated and dried sample was placed in an evaporating dish in a furnace (equipped with a circulating fan) at 80°C. The temperature was raised at the rate of about 50°C per hour to about 500°C and that temperature was held for one hour to assure oxide formation and then the temperature was allowed to return at furnace rate.

When the temperature had dropped to below 300°C, the sample was removed from the furnace and placed in a desiccator and the pressure reduced to 1½ inches of Hg or less. The vacuum was then broken with dry nitrogen. When the sample reached room temperature, it was weighed, placed in a glass jar, covered with water, and sealed. It should be noted that the firing temperature and time will depend on the coating used. For the zirconia coating, a temperature of about 500°C for about one hour is preferred although temperatures between about 200° and 800°C for one or more hours may be used, depending on the coating, e.g., when the aqueous zirconia solution is used, a temperature as low as about 200°C for about one hour is effective. However, temperatures above 800°C tend to result in poor coatings.

(E) Silanization (e.g., for About 2 g Quantities of Carrier): The desired quantity of zirconia coated CPG is weighed and placed in a large beaker. Ten milliliters

of 10% (by volume) aqueous silane, γ-aminopropyltriethoxysilane (A-1100, Union Carbide) is added to the coated glass and the pH of the mixture immediately adjusted to between 1.0 to 4.0, the optimum being around 3.5, using 6 N HCl. The reaction is carried out at a temperature of about 70°C for 3.5 hours, with occasional manual stirring. After the completion of the reaction, the glass is washed with distilled water, dried at 100° to 125°C and stored for use.

(F) Coupling Enzyme to Silanized Coated Carrier: Coupling via Glutaraldehyde Reaction — Ten ml of 2.5% by weight glutaraldehyde solution (in 0.1 M Na_2HPO_4 buffer solution at pH 7.0) is added per gram of silanized glass. The pH of the mixture is adjusted to 7.0 ± 0.2 with dilute HCl or NaOH. The reaction mixture is placed in a vacuum desiccator to remove trapped air bubbles and allowed to react at room temperature (25°C) for two to three hours.

At the completion of the reaction the excess glutaraldehyde is removed by rinsing several times with distilled water. After five or more rinses the excess water is filtered through filter paper (e.g., Whatman #41). The glass should immediately be subjected to the final enzyme coupling reaction.

Enzyme solution, dissolved in a minimum amount of water, containing 100 mg protein per gram of total carrier, is added to the above coated, silanized porous glass. For glucoamylase enzyme, the protein solution is prepared in 0.1 M Na_2HPO_4 buffer at pH 7.0, while gaseous isomerase enzyme solution is prepared in 0.05% $NaHCO_3$ buffer at pH 7.25.

Any other enzyme can be coupled at pH 7.0 to 8.0. Specific directions for coupling other enzymes may be found in U.S. Patent 3,519,538. The reaction is carried out at 5°C for three hours at a pH of 7.0 and 7.25, for glucoamylase and glucose isomerase, respectively. The pH adjustments are made with either dilute NaOH or HCl. After the completion of the coupling reaction, the glass is filtered and washed several times with distilled water. The immobilized enzyme is then stored under refrigeration in an appropriate buffer.

Coupling via Diazotization Procedure — Both glucoamylase and glucose isomerase may be coupled via azo linkage to the silanized coated carrier. The diazotization procedure is preferably carried out at 0°C in an ice bath. To 1 g of silanized, coated, porous glass add 10 ml of 2 N HCl. Next add 0.25 g solid $NaNO_2$. Place the entire reaction mixture in a desiccator and attach it to an aspirator to remove air and gas bubbles in the particles. Continue diazotization in aspirator for about 20 minutes. The desiccator is preferably packed in ice.

The product is then removed, filtered on a Büchner funnel and washed with ice cold water containing 1% sulfamic acid. Place the diazotized glass in the enzyme solution, using as little liquid as possible (slurry), e.g., the enzyme (100 mg/g of carrier) is dissolved in a minimum amount of water.

Previously adjust the enzyme solution to pH 8.0 to 9.0 and maintain the pH by addition of NaOH. Using a buffer at 0.05 M such as $NaHCO_3$ or Tris may aid in pH control. The reaction is continued for at least 60 minutes. Samples of enzyme may be withdrawn at intervals for protein determination to determine when maximum coupling rate has decreased (usually within 60 minutes). Continued reaction only adds additional enzyme (15 to 20%), but not necessarily additional activity. The product was washed with distilled water and stored in a closed container as taken from filter (usually 50 to 70% water by weight).

Enzyme composites having carriers which were treated were studied under conditions of catalysis. The porous glass carriers which had been coated with ZrO_2 coatings, and then silanized, and coupled with glycoamylase by azo linkages, were used in starch conversion studies, the results of which are summarized and compared with control composites in the table. From the table it can be seen that the coatings of ZrO_2 on carriers used in glucoamylase composites resulted in data from which calculated half-lives of 645 days were achieved.

These significant improvements in the half-lives of immobilized enzyme composites make possible the economical use of enzymes coupled to inorganic carriers since the increased half-life, in turn, greatly increases the value of an immobilized enzyme over free enzymes.

Glycoamylase was also coupled to alumina by known means and the resulting composites were used in similar studies which indicated a half-life of 70 days for the alumina-silane-enzyme composite (see table). The relatively high half-life found for the alumina carriers confirms the importance of having a carrier of low water solubility since it is known that alumina is less soluble than porous glass.

Comparison of Immobilized Glucoamylase on Controlled Pore Glass and Other Carriers

Carrier experiment	E_0 (units)	Temp. (° C.)	Volume (ml.)	Time (hrs.)	Decay constant (hrs.)$^{-1}$	Efficiency percent	$t_{1/2}$ (days)	Percent loss	Reactor No.
1-CPG[1]	57,471	60	16,815	528	3.35×10^{-3}	4.7	8.5	91	5
2-CPG	49,042	60	12,375	504	6.3×10^{-3}	1.0	5	96	8
3-CPG	19,923	45	10,150	552	1.33×10^{-3}	18.5	21.2	53	23
4-CPG-Zr[2]	16,245	45	76,000	504	4.5×10^{-5}	51	(645)	5	33
5-CPG-Zr	16,245	45	76,000	504	4.5×10^{-5}	51	(645)	5	34
6-Alumina[3]	4,170	45	2,400	336	4.1×10^{-4}	25	70	16	13a

[1] CPG represents porous 96% silica glass (controlled pore glass).
[2] CPG-Zr represents controlled pore glass-Zr coated.
[3] Neutral alumina 200 mesh.

NOTE.—Numbers in parenthesis are calculated values based on indicated hours of operation and initial and final activity.

Inert Carrier Coated with Polymer with Free Aldehyde Groups

E. Van Leemputten and M. Horisberger; U.S. Patent 3,821,083; June 28, 1974; assigned to Societe D'Assistance Technique Pour Produits Nestle SA, Switzerland have reported a process for the preparation of a water-insoluble enzyme product which comprises forming a layer of a water-insoluble polymer having free aldehyde groups on an inert carrier and thereafter reacting an enzyme with the free aldehyde groups of the polymer.

The term "inert carrier" is used to designate a water-insoluble substance in any form such as a membrane, a tube, granules or beads which does not react directly with the enzyme. Particularly suitable carriers include ceramic materials, glass, silica, silicates and metal oxides.

By "free aldehyde groups" are meant aldehyde groups carried by the polymer which are free to form chemical bonds. The enzymes used carry functional groups capable of reacting with the free aldehyde groups of the polymer, for example, free amino groups present in residues of amino acids such as lysine. Papain and trypsin are examples of preferred enzymes.

Polymer layers carrying free aldehyde groups may be formed on the inert carrier by various methods. In one preferred procedure, the polymer is dissolved in a solvent and the carrier contacted with the solution, as by immersion, and the solvent eliminated, e.g., by distillation and washing. The polymer may conveniently be polyacrolein, and crushed porcelain is a suitable carrier.

Polyacrolein comprises 10 to 70% of free aldehyde groups, which means that the number of available aldehyde groups corresponds to 10 to 70% of the number of molecules of monomer used. Although on deposition on the carrier it is only the free aldehyde groups at the surface of the polymer layer that are available for reaction with the enzyme, their number is sufficient to ensure adequate bonding.

Alternatively, a coating of a water-insoluble polymer which is soluble in common organic solvents and has reactive groups capable of being transformed into free aldehyde groups is deposited on an inert carrier. This deposition may be carried out by immersion of the inert carrier in the organic solution, followed by elimination of the solvent by distillation and washing. Free aldehyde groups are then regenerated on the polymer coating the inert carrier.

For example, a coating of polyacrolein butyl- or methylacetal in solution in a solvent such as ethyl acetate may be deposited on an inert carrier such as glass beads or crushed porcelain. After elimination of the solvent the regeneration of the aldehyde groups of the polymer is carried out by acid hydrolysis of the acetal, for example, in a column packed with the inert carrier coated with polymer.

According to a further variant, a polymerizable substance, such as acrolein, having at least one aldehyde function, which on polymerization forms a water-insoluble product having free aldehyde groups may be deposited on the inert carrier. Polymerization of this substance is then induced and the resulting carrier is recovered.

Irrespective of the method of preparation of the carrier, the reaction with the enzyme is conveniently effected by immersing the carrier in a solution of the enzyme or by passing the enzyme solution through a column packed with the carrier, and it has been found that the reaction parameters (quantities, time, temperatures, pH) may be varied within broad limits. Furthermore, the medium does not exercise a notable influence on the bonding reaction but, of course, it should be substantially inert with respect to the enzyme and carrier.

The quantities of enzyme and carrier used, as well as the concentration of the solution containing the enzyme, have no critical limits. The quantities and concentration may be chosen having regard to the nature of the reactants in order to obtain an optimum bonding yield.

As indicated, the pH and temperature may be chosen within broad limits, provided that the properties of the enzyme are not adversely affected. Although it is preferred to carry out the reaction at a pH of 7 to 8, desirably in a buffer solution, the bonding of the enzyme to the polymer may be carried out in the pH range extending from 1 to 11. Also, while ambient temperatures are preferred, the reaction temperature may be higher or lower, the upper limit being the inactivation temperature of the enzyme.

Reaction times may be chosen between a few minutes and several hours; most satisfactory results have been obtained with times of 3 to 5 hours. Upon completion of the reaction the enzyme product may be recovered by simple means such as filtration, decantation or centrifugation, followed by washing (distilled water, buffer solution). The quantity of enzyme bonded may be obtained by comparing the enzyme activity of the product, on an appropriate substrate, with the activity of the soluble enzyme on the same substrate.

Example 1: Polyacrolein is obtained by polymerization of the monomer using as initiator a redox system of hydrogen peroxide-sodium nitrite, and a 1% solution of the polymer in tetrahydrofurfuryl alcohol is prepared.

50 ml of porous crushed porcelain, the granules of which have an average size of 2 to 3 mm, are immersed in this solution and the tetrahydrofurfuryl alcohol is distilled under vacuum. The porcelain is then carefully washed with distilled water. During this washing, the last traces of alcohol pass into solution in the water and are eliminated. The wash water may also entrain a certain amount of polyacrolein, but this quantity is insufficient however to adversely affect the subsequent bonding of the enzyme.

The washed coated porcelain is then immersed for 8 hours at ambient temperature in a buffer solution of 0.05 M sodium barbital hydrochloride, at pH 8.0, containing 100 mg of papain per 25 ml of solution. The reaction may also be carried out by cycling the solution through a column packed with the carrier.

The resulting product is then washed with distilled water, then with a buffer solution at pH 7.2, containing, per 500 ml of solution, 0.605 g of cysteine hydrochloride, 6.75 g of potassium phosphate and 10 ml of a decinormal solution of the disodium salt of ethylene diamine tetraacetic acid.

The estimation of the enzyme activity of the product shows that 13 mg of papain are bonded to 50 ml of porcelain.

Example 2: Polyacrolein methylacetal is prepared by a method similar to that described by Weygand-Hilgetag in *Organisch-Chemische Experimentierkunst,* page 455 (1964).

5.6 g of polyacrolein having 6.2% of reducing groups are suspended in 80 ml of methanol. 13 ml of trimethyl orthoformate are then added, followed by 0.05 ml of concentrated hydrochloric acid and the mixture is heated until the polyacrolein has completely dissolved. The solution is then neutralized with a 30% sodium hydroxide solution and concentrated by evaporation of the methanol under reduced pressure.

The resulting syrup is taken up in 300 ml of chloroform, washed twice with a saturated aqueous sodium bicarbonate solution, and finally three times with water. The organic phase is dried on anhydrous sodium sulfate, filtered and evaporated. The resulting polyacrolein methylacetal is dissolved in 30 ml of ethyl acetate.

50 ml of crushed porcelain are immersed in 100 ml of ethyl acetate to which are added 5 ml of the polyacrolein methylacetal solution in ethyl acetate. The solvent is then eliminated by evaporation under reduced pressure followed by heating at 70°C under vacuum for 30 minutes.

The free aldehyde groups of the polymer are then regenerated by immersing the porcelain coated with polyacrolein acetal for 8 hours, at ambient temperature, in a normal solution of hydrochloric acid. Regeneration may also be carried out in a column.

The experimental determination of the presence of free aldehyde groups regenerated by acid hydrolysis of polyacrolein methylacetal, as well as their quantitative estimation, were performed in parallel on a polymer in powdered form. The estimation of reducing groups according to the Park and Johnson method, adapted to insoluble materials, as described by J.S. Thompson and G.D. Shockman in *Analytical Biochemistry*, 22, 260-268 (1968), showed a number of free aldehyde groups equal to 9% of the number of molecules of monomer used.

Even though only the superficial aldehyde groups are available for bonding, the results given below show that their number is sufficient to ensure adequate bonding of the enzyme.

The coated porcelain, after acid hydrolysis of the polyacrolein methylacetal, is treated in a column by circulation of 20 ml of a buffer solution of 0.05 M sodium barbital hydrochloride at pH 8.0, containing 200 mg of trypsin. This treatment is performed for 5 hours at ambient temperature, with recycling of the solution. The estimation of the enzyme activity of the final product shows that 10 mg of trypsin are bonded to 50 ml of porcelain.

Example 3: Polyacrolein n-butylacetal is obtained by suspending 10 g of polyacrolein in 200 ml of n-butanol in the presence of 1 g of p-toluene sulfonic acid as acetylization catalyst. After heating at boiling point for 5 minutes the polyacrolein is dissolved. The solution is concentrated by evaporation under vacuum, taken up in chloroform (or another solvent such as benzene or toluene) washed twice with a saturated aqueous solution of sodium bicarbonate, and finally three times with water.

The organic phase is dried on anhydrous sodium sulfate and evaporated. A 10% solution of the resulting polyacrolein n-butylacetal is prepared in ethyl acetate. 80 g of glass beads, 0.3 mm in diameter, are immersed in 100 ml of ethyl acetate to which is added 1 ml of the 10% polyacrolein n-butylacetal solution in ethyl acetate. The solvent is eliminated by evaporation under vacuum and heating at 70°C under vacuum for 30 minutes.

The free aldehyde groups of the polymer are then regenerated by immersing the coated glass beads in a normal solution of hydrochloric acid at ambient temperature for 8 hours or by using a column.

The resulting product is then placed in a glass spiral or a column, washed with distilled water and treated by circulating 30 ml of a buffer solution of 0.1 M sodium barbital hydrochloride containing 100 mg of papain. This treatment is carried out for one hour at ambient temperature with a throughput of solution of 1.4 ml per minute. The resulting product, after washing at 42°C with a buffer solution of cysteine hydrochloride at pH 7.2, contains 5.6 mg of bonded papain.

COPOLYMERIZATION

Enzyme-Epoxy Compound Polymerized with Olefinically Unsaturated Monomer

The processes of *K. Beaucamp, H.U. Bergmeyer, K.-H. Botsch, D. Jaworek and M. Nelboeck-Hochstetter; U.S. Patent 3,806,417; April 23, 1974; assigned to Boehringer Mannheim GmbH, Germany* provides an especially mild process for binding proteins to insoluble carriers, which gives products which not only have a high activity and activity yield but which are also, above all, so stable that they can be used for technical purposes.

A process has been described for the preparation of carrier-bound proteins, where a protein is reacted in aqueous solution with a compound containing at least one epoxy group and at least one further functional group capable of producing a bond with a carrier substance, whereafter this further functional group is reacted with a carrier substance.

The process results in the protein forming a covalent bond with the epoxy group-containing compound, with opening of the epoxy groups. The intermediate product thus obtained is, without further isolation, connected with an appropriate carrier by reacting the further functional group with the carrier substance.

The further functional group is preferably one which can add on to or condense with the actual carrier substance. When epoxy compounds are used which contain a group capable of condensation, care is to be taken that, during the condensation, no substances are split off which disadvantageously influence the activity of the bound protein. In order to ascertain whether the use of particular group capable of condensation results in the splitting off of a product which impairs the activity of the initially used enzyme, can easily be determined for each particular protein to be bound, by means of a few simple preliminary experiments.

However, general predictions regarding the suitability of particular groups cannot be made because the various active proteins have greatly differing sensitivities. For example, the splitting off of halides can lead to a loss of activity of many sensitive proteins, whereas other active proteins are not disadvantageously influenced. The connecting of the intermediate product with the carrier material can take place in an especially mild manner by polymerizing it into the carrier substance. Therefore, according to the process, there is used an epoxy compound which contains at least one copolymerizable double bond as further functional group which can form a bond with a carrier material.

The carrier substances can be any water-insoluble solid substances which via the further functional group of the epoxy compound, can be connected therewith in aqueous solution under mild conditions. Carrier substances are preferably used which are hydrophilic, easily swellable and substantially free of charges and also stable towards micro-organisms. The carrier substance can, as such, be introduced into the aqueous solution for the production of the bond with the intermediate product but preferably the carrier substance is produced in the aqueous solution itself by polymerization of water-soluble monomers.

The reaction of the protein with the epoxy compound can either take place (a) in the presence of the polymerizable monomer(s), the polymerization thereby being achieved with polymerizing of the epoxy compound-protein intermediate product, or (b) the polymerizable monomers or monomer mixture is added to the solution after the reaction between the protein and epoxy compound and then the polymerization is initiated.

As monomers of the process, there can be used those water-soluble compounds which can undergo polyaddition or polycondensation. Monomers capable of polyaddition are preferred and especially those monomers which contain at least one olefinic double bond. In this case, the further functional group of the epoxy compound is preferably also a copolymerizable double bond.

Preferred epoxy compounds for the process have the general formula:

$$R_1-(CH_2)_n-CH\overset{\displaystyle O}{\overbrace{\qquad}}CHR_2$$

wherein

R_1 is a mono- or polyunsaturated alkenyl radical, an alkenyloxy radical or an unsaturated acyl, i.e., alkanoyl, the CO group of which is preferably conjugated with a carbon-carbon double bond;

n is 0, 1 or 2; and

R_2 is a hydrogen atom or a lower alkyl radical.

Lower alkyl radical refers here to a straight or branched radical containing up to 4 carbon atoms. The substituent R_1 preferably contains 2 to 6 carbon atoms but longer radicals of, e.g., up to 10 carbon atoms, can also be used so long as the water-solubility of the epoxy compound is not too strongly reduced. Preferred epoxy compounds can be derived from 2,3-epoxypropanol, in which a hydrogen is replaced by a lower alkyl radical, which is connected with a further olefinically-unsaturated compound with ester or ether formation.

Depending upon the desired consistency of the end product, cross-linking compounds can also be added to the monomer which contain more than one polymerizable group. Examples of such cross linkers include N,N'-methylenebisacrylamide and ethylene diacrylate. These are preferred when working in aqueous solution. If the polymerization is carried out in a heterogeneous phase, i.e., a suspension polymerization, water-insoluble cross-linkers, for example, divinylbenzene and ethylene dimethacrylate, can also be used.

Numerous other cross-linkers are also known and the appropriate choice to be made in a particular case lies within the scope of the expert's ability. It is also possible subsequently to cross-link carrier-bound proteins obtained by the process, the carrier of which is not cross-linked.

If no cross-linker is used, then carrier materials are obtained which are soluble or thermoplastic. Carrying out the process in this way gives spinnable or extrudable solutions from which the carrier-bound proteins can be obtained, for example, in filament form or in the form of foils, by known methods.

Filaments or foils of this kind, with active protein covalently bound, can be stretched, spun and worked up to other products according to the known methods of synthetic resin technology and contain the bound active protein and can be used for purposes in which these forms offer especial advantages, for example, for the production of enzymatically-active sieves, fabrics, implantable filaments and the like.

For spinning from aqueous solution, there can, for example, be used the vacuum spinning process in which the solution is forced thorugh a spinneret into a vacuum. This can take place under the conditions of a lyophilization, which can be tolerated by most active proteins, without loss of activity.

It is also important, for the process that the protein is first reacted with the epoxy compound, whereafter the intermediate product obtained is fixed on to the carrier substance. The reaction of the protein with the epoxy compound normally needs no special measures. It is usually sufficient to bring the protein and epoxy compound together at ambient temperature in aqueous solution.

It is hereby expedient to work in a buffered aqueous solution, the pH value of which is appropriate for the protein in question. The period of reaction between protein and epoxy compound depends upon the particular substances used but, in general, is between about 5 minutes and 1 hour. Longer or shorter incubation times can, however, also be expedient from case to case.

The reaction of protein and epoxy compound can, as mentioned, be carried out in the presence of the carrier or of the starting products for the carrier. In the latter case, the polymerization reaction is expediently initiated, after the formation of the preproduct, by the addition of an initiator. As initiators, there can be used the initiators and catalysts conventionally employed in polymer chemistry insofar as they do not disadvantageously influence the activity of the protein. As initiators or catalysts, there can, for example, be used, in the case of olefinically-unsaturated monomers or prepolymers, inorganic or organic peroxides, azo compounds and the like. In addition, reaction accelerators, such as amines and the like, can be used. When using acrylic acid or methacrylic acid derivatives as monomers, the use of an initiator combination of a peroxydisulfate and an amine, such as 3-dimethylaminopropionitrile, has proved to be especially useful. When using this initiator combination, it is expedient to work under an inert atmosphere, for example, under nitrogen.

When the process products are directly formed as insoluble materials, they can be isolated by simple filtering, followed by washing. If the carrier is not cross-linked and remains in solution, the solvent can be removed in the usual way, for example, as mentioned above by vacuum spinning.

Example 1: Starting Substances —

> 0.1 g glucose oxidase (GOD), 220 u/mg, acrylamide,
> acrylic acid-(2,3-epoxypropyl ester),
> N,N'-methylenebisacrylamide,
> 3-dimethylaminopropionitrile,
> ammonium peroxydisulfate

Method: 0.1 g GOD were dissolved at 30°C in 2 ml 0.5 M triethanolamine buffer (pH 8.0) under an atmosphere of nitrogen mixed with 0.25 ml acrylic acid-(2,3-epoxypropyl ester) and stirred for 30 minutes. Subsequently, the reaction mixture was cooled to +10°C and mixed with 3 g acrylamide and 0.1 g N,N'-methylenebisacrylamide dissolved in 18 ml distilled water.

Under an atmosphere of nitrogen, after about 15 minutes, the polymerization was started with 0.5 ml 5% 3-dimethylaminopropionitrile and 3 ml 5% ammonium peroxydisulfate. The solution first became viscous after about 5 minutes and then further polymerized to give a yellow, clear and stiff gel. The polymerization block obtained was then left to stand for about 18 hours in a refrigerator. The polymerizate was subsequently forced through a metal sieve of 0.5 mm mesh size and washed with about 2,000 ml distilled water.

The gel particles were then filled into a column of 2 cm internal diameter and washed out with 0.2 M potassium phosphate buffer (pH 7.5) to remove adsorptively bound GOD until no enzyme activity was detected in the eluate. 10% of the originally used enzyme activity was detected in the eluate. The enzymatic activity of the carrier-bound enzyme, obtained as a lyophilizate, was 500 u/g.

Example 2: Starting Substances —

> uricase, 4.5 u/mg, dissolved in 50% glycerol,
> 0.05 M glycine and 0.13 M sodium carbonate,
> acrylamide,
> acrylic acid-(2,3-epoxypropyl ester),
> N,N'-methylenebisacrylamide,
> 3-dimethylaminopropionitrile,
> ammonium peroxydisulfate

Method: 10 mg uricase were dialyzed at 4°C for about 4 hours against 1 liter 0.01 M glycine buffer (pH 10). The dialyzate was mixed with 1 ml 1 M triethanolamine buffer (pH 8.0) and was incubated at 10°C, under an atmosphere of nitrogen, in the presence of 0.1 ml acrylic acid (2,3-epoxypropyl ester) and stirred for 30 min. Working up took place as described in Example 1. The activity of the lyophilizate obtained was 3.5 μ/g.

Example 3: Starting Substances —

> 140 μ/mg hexokinase,
> acrylamide,
> acrylic acid (2,3-epoxypropyl ester),
> N,N'-methylenebisacrylamide,
> 3-dimethylaminopropionitrile,
> ammonium peroxydisulfate

Method: 20 mg hexokinase crystal suspension were dissolved in 5 ml water and dialyzed at 4°C against 1 liter 0.01 M triethanolamine buffer (pH 8.0). The reaction mixture was incubated at 10°C under an atmosphere of nitrogen in the presence of acrylic acid (2,3-epoxypropyl ester) and, after 30 min, was cooled to 10°C. The enzyme solution was then mixed with 3 g acrylamide and 0.1 g N,N'-methylenebisacrylamide, dissolved in 20 ml distilled water. The further working up took place as described in Example 1. The activity of the lyophylizate obtained was 12 μ/g.

Example 4: Starting Substances —

> 220 μ/mg glucose oxidase (GOD),
> 1,2-epoxybut-3-ene,
> acrylamide,
> N,N'-methylenebisacrylamide,
> 3-dimethylaminopropionitrile,
> ammonium peroxydisulfate

Method: 100 mg GOD were dissolved at 30°C, under an atmosphere of nitrogen, in 2 ml 0.5 M triethanolamine buffer (pH 8.0) and then incubated for 30 min with 0.05 ml 1,2-epoxybut-3-ene. Subsequently, the reaction mixture was cooled to 10°C and then mixed with 3 g acrylamide and 0.125 g N,N'methylenebis-acrylamide, dissolved in 18 ml bidistilled water. The reaction mixture was then further worked up in the manner described in Example 1. 37% of the original enzyme activity was detected in the eluate. The lyophilizate obtained had an activity of 540 μ/g.

Copolymerized Acylase and Anhydrides of N-Carboxy-α-Amino Acids

Acylase causes asymmetrical removal of acyl groups from acyl-DL-amino acids, and racemic amino acids can be resolved into their optically active enantiomorphs by acylating the amino radical of the racemate, and by selectively hydrolyzing one of the acylated enantiomorphs by means of acylase. Acylase can be isolated from fungi, bacteria and from various animal tissues. The isolation of acylase from its natural sources is a complex and costly procedure. The recovered acylase is water-soluble and sensitive to environmental factors which tend to deactivate the enzyme.

J. Kirimura and T. Yoshida; U.S. Patent 3,243,356; March 29, 1966; assigned to Ajinomoto Co., Inc., Japan have found that acylase can be made insoluble in water and that its stability can be greatly increased without loss of enzyme function by copolymerizing acylase with anhydrides of certain N-carboxy-α-amino acids or derivatives to produce polypeptidyl derivatives of acylase.

The copolymerization reaction may be carried out in an aqueous system buffered to a pH value between 5 and 9 at which no adverse effect on acylase activity is noticeable. The polypeptidyl derivatives of acylase obtained by the reaction is practically insoluble in water. It has approximately the same enzyme effect as equivalent amounts of unmodified acylase, and it can be stored without loss of activity over extended periods.

When the modified insoluble acylase is brought into contact with acyl-DL-amino acids, the racemate is asymmetrically hydrolyzed in the same manner as with soluble acylase but the insoluble enzyme can be recovered from the hydrolyzation mixture. It does not contaminate the product and it can be used again for the hydrolysis of another batch of acyl-DL-amino acid.

Typical amino acids whose N-carboxy anhydride derivatives may be employed include glycine, alanine, α-amino-n-butyric acid, valine, leucine, isoleucine, phenylalanine, the β-methyl, β-ethyl, and β-benzyl esters of aspartic acid, and the corresponding γ-esters of glutamic acid. The modification reaction can be performed in a homogeneous liquid medium, or in a heterogeneous medium. The reaction products are practically insoluble in water.

The N-carboxy anhydrides of proline, serine, or threonine react with acylase to form polypeptidyl derivatives but these modified acylase products are water-soluble and, therefore, no more useful than the unmodified acylase.

The asymmetric removal of acyl radicals from N-acyl-DL-amino acids can be carried out in continuous operation. A column is packed with the modified water-insoluble acylase, and a solution of the N-acyl-DL-amino acid is passed through the column. The optically active amino acid is found in the column effluent, and is readily recovered.

The column packing preferably includes an inert extender or carrier in addition to the modified acylase to increase the available surface of the latter and to control the rate of flow in the column. Powdered polyvinyl chloride, polyethylene, polystyrene and purified diatomaceous earth are typical of the many inert and insoluble organic and inorganic extenders or carriers which may be employed.

As long as the rate of flow does not exceed the capacity of the column, the concentration of the optically active amino acid in the effluent is approximately proportional to the concentration of the N-acyl-DL-amino acid in the material fed to the column.

Example 1: Purified acylase was reacted with N-carboxy-γ-methylglutamic acid anhydride, which had been prepared from γ-methylglutamic acid, in a $\frac{1}{15}$ M acetate buffer solution at pH 7.2. The water-insoluble modified acylase recovered from the reaction mixture was rapidly powdered in a chilled mill.

Polyvinyl chloride powder was separately washed with water to remove all ex-
tractable material. 40.0 g of the polyvinyl chloride powder and 1.0 g of the
modified acylase powder were mixed. The mixture was moistened with 0.05 M
barbiturate buffer solution at pH 8.0 and packed into a column 2 cm in diameter
and 30 cm high. 200 ml of the buffer solution were passed through the column
at the rate of 60 ml/hr for further washing and stabilizing the acylase.

Ten liters of an aqueous solution containing 1,000 g N-acetyl-DL-valine and 10^{-3}
mol cobaltous chloride were adjusted to pH 7.8 with sodium hydroxide and
percolated through the column at room temperature at the rate of 60 ml/hr.
The top surfaces of the percolated solution and of the effluent from the column
were covered with toluene to prevent bacterial attack.

The pH of the collected effluent was adjusted to pH 5.0, and its volume was reduced to one-fifth by evaporation in a vacuum. The residue was diluted with
99.5% ethanol to the original volume. The mixture was kept in a refrigerator
overnight. L-valine crystallized, and the crystals were separated from the mother
liquor by filtration.

They were recrystallized from water and ethanol. The L-valine obtained was
white and flaky. It weighed 298 g, and its specific rotation was $[\alpha]_D^{23}$ +27.2
(c = 2%, in 5 N HCl). The yield was 81% of that theoretically possible.

Example 2: A solution of purified acylase, 450 ml, was adjusted to pH 7.2 and
sodium acetate and cobaltous chloride were added in amounts of mol/15/l and
10^{-4} mol/l respectively. To this acylase solution, a solution of N-carboxy-γ-methyl-
L-glutamic acid in dioxane (3.0 g/90 ml) was added with agitation drop by drop
while the reaction medium was kept at 2°C by ice-cooling.

Stirring was continued for 30 min, after the dioxane solution had been added,
and the reaction mixture was thereafter kept in an icebox for 20 hr at 2° to 4°C.
The precipitate formed in this period was collected by centrifuging at 0°C, and
washed repeatedly with a solution of sodium acetate which contained cobaltous
chloride. When ninhydrin positive material could no longer be detected in the
washings, the residue was collected and lyophilized.

The water-insoluble poly-γ-methyl-L-glutamyl acylase thus obtained was a pale
yellow granular material whose yield was 1.013 g. Its content of original acylase
was calculated as 11.6% from the amount of unreacted N-carboxy-γ-methylgluta-
mate which was recovered from the reaction mixture. The acylase activity of
the water-insoluble acylase was assayed as 12.56 micromols/mg/30 min for N-
acetyl-DL-valine.

CROSS-LINKING

POLYMER SURFACE AS A SUPPORT

J.H. Reynolds; U.S. Patent 3,705,084; December 5, 1972; assigned to Monsanto Company has described immobilized enzymes which are adsorbed to polymeric material and adapted to be used in a macroporous core of an enzymatic reactor. The active enzyme adsorbed on the polymeric surface is further immobilized by cross-linking in place with a cross-linking agent such as dialdehyde, monomeric polyisocyanate, bisamidoester, disulfonyl halide, and the like. The macroporous reactor core can be made of acrylic type or polyamide type macromolecules, polyurethane, and the like, thereby providing an adsorption-promoting surface, or it can be made of an inert material, such as a foamed open cell, polyolefin or the like, and then the surfaces thereof coated so as to provide the desired polymeric surface thereon.

Such immobilized enzymes, when enclosed in a holder or column fitted with inlet and outlet ports through which the substrates are passed and attached to appropriate pumping, filtering, and monitoring equipment, comprise an enzyme reactor system, the relatively high porosity of the reactor core allowing high flow rates through the reactor.

Enzyme reactors are particularly suitable for applications in cheese manufacture for clotting of milk, in the production of glucose from starch, in sweetening of crude syrups, in chill-proofing of beer, in synthesis of organic intermediates, in production of flavor enhancers such as nucleotides, in specific waste treatments, for the removal of undesirable polysaccharides from soy milks, for the determination of blood glucose and other components of physiological fluids, and the like.

The polymeric surface suitable for the attachment of enzymes can be made of a water-insoluble polymer having nitrilo ($\equiv$N), acid amido ($-CONH_2$) or ureido ($-NHCONH_2$) groups which promote enzyme adsorption. Typical of such polymers is an acrylic resin such as the polymers or copolymers of acrylic acid, methacrylic acid, esters of the foregoing acids or acrylonitrile, for example, poly-acrylamide and polyacrylonitrile. Also suitable are polyamide macromolecules such as a polylactam produced by the polymerization of lactam monomers of the formula

$$R' \underset{NH}{\overset{\frown}{\bigg\langle}} C=O$$

wherein R' is an alkylene group having 3 to about 12 carbon atoms, preferably 5 to 12 carbon atoms. Suitable lactam monomers are epsilon-caprolactam, alpha-pyrrolidone, piperidone, valerolactam, methylcyclohexanone oximes, capryllactam, cyclodecanone isoxime, lauryllactam, and the like. Typical polyamide resins are polycaprolactam (nylon 6), polyhexamethyleneadipamide (nylon 6,6) and the like. Also contemplated are polyamides proteinaceous in nature such as collagen, albumins, and the like.

Another type of polymer that can be advantageously utilized for this purpose is a polyurethane resin which is produced by the reaction of polyfunctional iso-cyanates with organic compounds containing active hydrogen atoms such as al-cohols (including phenols), carboxylic acids, polyester adducts derived from the condensation of polyols with polycarboxylic acids and anhydrides, and the like, i.e., with organic compounds such as polyesters or polyethers which contain at least two free hydroxyl groups, in the presence of various adjuvants well known in the art such as catalysts, emulsifiers, foaming or blowing agents, or the like. Polyacrylonitrile, polycaprolactam and polyurethane are the preferred polymeric materials.

Example 1: Alkaline Protease Reactor — A polyethylene disk 38 mm in diameter and $\frac{1}{8}$ inch thick was cut from a sheet of porous polyethylene and was permeated with a 10 weight percent solution of polyacrylonitrile in N-methylpyrrolidinone at 50°C. Thereafter excess polymer solution was removed from the outer sur-faces of the disk, and the disk was placed in a suitable holder. Distilled water was then pumped through the disk to precipitate the polyacrylonitrile on the internal disk surfaces and to remove the N-methylpyrrolidinone.

An aqueous enzyme solution containing alkaline protease (subtilisin Carlsberg) (0.1% by weight) and glutaraldehyde (3% by weight) and having a pH of 6 was prepared and pumped repeatedly through the disk for about 3 hours. There-after the disk was washed with tris(hydroxymethyl)aminomethane buffer solution (pH 8.0) to which potassium chloride (1 M) has been added to wash out unbound enzyme.

The assay procedure comprised adding 0.1% denatured lysozyme to tris(hydroxy-methyl)aminomethane buffer (pH 8.0) solution and pumping the resulting solution through the disk at a rate of 8.0 ml/min. The effluent was collected in test tubes containing aqueous 5 weight percent trichloroacetic acid solution, filtered to re-move any precipitate that may be present and measuring the absorbence of the filtrate in a spectrophotometer at a wavelength of 280 mμ. Esterase activity of the subtilisin reactor was determined using solutions of benzoylarginine ethyl ester. An enzymatically active reactor core was produced.

Example 2: Subtilisin Reactor — Long term protease activity of a subtilisin reactor prepared as in Example 1 towards casein was demonstrated. A 0.1 weight percent solution of casein was passed through the reactor for three weeks. At the end of that time it had lost only 10% of its original activity. Similar results were obtained with benzoyl arginine ethyl ester as substrate.

Example 3: Chymotrypsin Reactor — A polyethylene disk 140 mm in diameter and ⅛ inch thick was cut from a sheet of porous polyethylene and was permeated with a 10 weight percent solution of polyacrylonitrile in N-methylpyrrolidinone at 50°C. Thereafter excess polymer solution was removed from the outer surfaces of the disk, and the disk was placed in a membrane holder. Distilled water was then pumped through the disk to precipitate the polyacrylonitrile on the disk surfaces and to remove the N-methylpyrrolidinone.

An aqueous enzyme solution containing α-chymotrypsin (0.1% by weight) and glutaraldehyde (3% by weight) and having a pH of 6.5 was prepared and pumped repeatedly through the disk for about 1 hour. Thereafter the disk was washed with tris(hydroxymethyl)aminomethane buffer (pH 8.0) solution to which potassium chloride (1 M) had been added to wash out unbound enzyme and then assayed. Esterase activity was demonstrated with acetyl tyrosine ethyl ester, and benzoyl tyrosine ethyl ester. Amidase activity was demonstrated with carbobenzoxy-tyrosine p-nitroanilide. Milk clotting activity was observed when a solution of 6% dry milk (nonfat) was passed through the reactor.

Example 4: Trypsin Reactor — A polyethylene disk 38 mm in diameter and ⅛ inch thick was cut from a sheet of porous polyethylene and was soaked in a 10 weight percent solution of nylon 6,6 in formic acid at 50°C. Thereafter excess polymer solution was removed from the outer surfaces of the disk, and the disk was placed in a membrane holder. Distilled water was then pumped through the disk to precipitate the nylon 6,6 on the disk surfaces and to remove the formic acid.

An aqueous enzyme solution containing trypsin (0.1% by weight) and glutaraldehyde (3% by weight) and having a pH of 6.5 was prepared and pumped repeatedly through the disk for one hour. Thereafter the disk was washed with tris(hydroxymethyl)aminomethane buffer (pH 8.0) solution to which potassium chloride (1 M) has been added to wash out unbound enzyme and then assayed. Enzyme activity was demonstrated with benzoyl arginine ethyl ester and tosyl arginine methyl ester, and amidase activity towards benzoyl arginine p-nitroanilide. Protease activity was observed with casein.

Example 5: Alkaline Phosphatase Reactor — A polyethylene disk 38 mm in diameter and ⅛ inch thick was cut from a sheet of porous polyethylene and was soaked in a 10 weight percent solution of polyacrylonitrile in N-methylpyrrolidinone at 50°C. Thereafter excess polymer solution was removed from the outer surfaces of the disk, and the disk was placed in a membrane holder. Distilled water was then pumped through the disk to precipitate the polyacrylonitrile on the disk surfaces and to remove the N-methylpyrrolidinone.

An aqueous enzyme solution containing alkaline phosphatase (chicken intestine) (0.1% by weight) was pumped repeatedly through the disk for about ½ hour.

The enzyme was then cross-linked with glutaraldehyde as in Example 4. There-after the disk was washed with glycine buffer solution to which potassium chlo-ride (1 M) had been added to wash out unbound enzyme and then assayed. The alkaline phosphatase reactor was assayed with o-carboxyphenyl phosphate and found to have substantial enzyme activity.

PHENOL-FORMALDEHYDE RESIN AS SUPPORT

Insolubilized but active enzymes are prepared from enzymes which are in a normal or native (soluble) state by reacting them with a phenol-formaldehyde resin and glutaraldehyde. The reaction is generally conducted in an aqueous medium, and preferably the resin and glutaraldehyde are sequentially reacted with the starting enzyme according to *A.C Olson and W.L. Stanley; U.S. Patent 3,767,531; October 23, 1973; assigned to U.S. Secretary of Agriculture.*

The insolubilized (immobilized) enzyme products are referred to by the term enzyme-RG, and specifically as lactase-RG, amylase-RG, etc. A primary advan-tage of the products is that their activity does not materially decrease during use. For instance, a product (lactase-RG) was used continuously for over 6 weeks with no loss of activity under conditions wherein the ratio of substrate to enzyme reaction product was greater than 100,000 to 1. Thus, the products have the advantage not only of being reusable, but also usable under conditions of con-tinuous operation for long periods of time and with large amounts of substrates.

Another advantage is that the phenol-formaldehyde resin not only contributes to insolubilization of the enzyme applied thereto, but also provides useful physical properties to the product. In particular, the resin acts as a support or carrier so that the insolubilized enzyme product forms a column through which water and other liquids can percolate readily. This is in sharp contrast to known insolu-bilized enzymes which are generally amorphous materials that cannot be used directly in a column because liquids will not flow therethrough. These known products require the addition of a carrier such as diatomaceous earth, crushed firebrick, or the like to provide a liquid-permeable mass.

Another advantage is that the products are afforded by simple procedures using readily-available reactants. No exotic chemicals or complicated procedures are required. Nonetheless, the products retain a significant and sufficient part of the activity of the starting enzyme. In some cases, the major part of the original activity is retained.

A further advantage is that useful products can be prepared from any enzyme source, including pure enzymes, enzyme concentrates, crude enzyme preparations, and even such substances as animal organs, plant parts, microbial cultures, and the like. Important in this regard is that application of the herein-described reactants causes most of the active enzyme to be selectively precipitated even where it is present in minute quantity, e.g., as little as 1 mg of active enzyme in association with gram quantities of inactive components. Accordingly, the process provides the means for preparing insolubilized products from enzymes which previously were difficult to insolubilize or which were never insolubilized. Another advantage lies in the precise control that one can exercise over the extent and direction of enzymic reactions. This results because of the solid nature of the products which allows specific amounts to be metered out to suit any particular situation. Another advantage is that enzymic reactions can be stopped at any desired time by simply separating the solution under treatment from the insolubilized enzyme, for example, by draining the solution away from the reaction system. Thus, no external forces, such as heat, acid, and the like which might be detrimental, need be applied to short-stop the reaction.

Preparation of the products involves reaction of the starting enzyme with the phenol-formaldehyde resin and glutaraldehyde. In a preferred procedure, the starting enzyme is first adsorbed on the resin and the resin-enzyme complex is then treated with glutaraldehyde. Alternatively, the resin may be first treated with glutaraldehyde and then with the enzyme. Water is advantageously used as the reaction medium. Temperatures employed are generally ambient (room) temperature, or somewhat lower or higher, i.e., the range about from $1°$ to $40°C$. Conventional operations such as mild stirring or shaking are applied to attain good contact between reactants. The products are segregated by the usual mechanical procedures such as filtration, centrifugation, or decanting. For best results, the aqueous medium is adjusted to the pH at which the enzyme in question is soluble.

In another procedure, the resin is first treated with a nonenzymatic protein (such as bovine serum albumin, soybean protein, egg albumin, or the like), then treated with glutaraldehyde, and finally with the enzyme to be insolubilized, whereby to yield a high specific activity product. The phenol-formaldehyde resins are commercially-available resins produced by the condensation polymerization of phenol, or a substituted phenol, with formaldehyde under acidic or basic conditions.

For use in the process, the phenol-formaldehyde resin should be in granular or bead form. If the resin is in large pieces, grinding is applied to reduce it to granular form. It is also desirable to apply a sieving operation to remove fine particles and oversize particles retaining those having a mesh size in the range about from 10 to 40 mesh.

The enzyme to be insolubilized is dissolved in distilled water. Where necessary, the pH of the water is adjusted by conventional methods to a level at which the enzyme is soluble. Appropriate pH's to use with any particular enzyme are described in the literature. In many cases a pH of about 3 to 7 is employed. It may further be noted that oftentimes the starting material already contains a buffer or other pH-adjusting agent so that when it is mixed with water the resulting dispersion will exhibit a pH at which the enzyme is most soluble. This is particularly the case with commercially-available enzyme preparations. It is obvious that in such cases there is no need to apply any pH adjustment.

Following preparation of the aqueous solution of the starting material, a mechanical separation step such as filtration or decantation can be applied to remove fillers, debris, or other undissolved material. Next, the aqueous dispersion of the starting enzyme is added to the resin granules. Generally, about 20 to 100 parts of resin per part of enzyme are used. The mixture is gently agitated by conventional means such as shaking, stirring, or the like while being held for about ¼ to 5 hours at a temperature of about 1° to 25°C in order to cause the enzyme to be adsorbed on the resin. Having adsorbed the enzyme on the resin, an aqueous solution of glutaraldehyde is added to the above suspension. The amount of glutaraldehyde is not critical. Usually, a large excess, e.g., 10 to 50 parts thereof per part of enzyme, is used; the unreacted residue is removed in a subsequent washing step.

The resulting mixture is held for a period of time to ensure formation of the enzyme-RG product. Usually, the holding is for a period of about 12 to 24 hours in a cold room at about 1° to 10°C. However, a somewhat shorter holding period can be realized if the holding is conducted at temperatures between 10° to 25°C. The product is then collected by filtration and rinsed several times with distilled water to remove excess reagents and other impurities. The so-prepared enzyme-RG is then ready for use.

Example 1: Preparation of Lactase-RG — The starting enzyme was a commercial product (Wallerstein lactase LP), the lactase content of which was estimated to be about 10%. The resin was a commercial phenol-formaldehyde resin (Duolite S-30).

The resin was sieved to obtain particles of 30 to 40 mesh. After washing the resin with distilled water, it was soaked overnight in 0.1 M (0.58%) sodium chloride and rewashed with distilled water. A solution of 228 mg of the lactase preparation in 20 ml of distilled water was added to 10 grams of wet, drained resin at 25°C. After stirring gently for 15 minutes, 3 ml of a 25% aqueous glutaraldehyde solution was added. The reaction mixture was left in the cold (5°C) for 16 hours and then rinsed several times with distilled water. The immobilized enzyme product exhibited 60% of the activity of the original enzyme.

Example 2: Lactase-RG Hydrolysis of Lactose — Lactase-RG prepared as described in Example 1 was packed into a 1.2 cm x 15 cm jacketed column over a bed of sand. The column was then washed several times with pH 4.5 (0.02 M) phosphate buffer. An aqueous lactose solution (3.0% in 0.02 M phosphate buffer) was then pumped through the column at varying flow rates. The column temperature was regulated by passing warm water through the column jacket. The extent of hydrolysis was determined by analyzing for glucose in the effluent. The results are summarized below.

Temperature, °C	30				45					55		
Flow, ml./min.	0.2	0.5	1	2	0.2	0.5	1	2	5	0.5	1	2
Percent hydrolysis	90	76	58	43	100	93	81	64	44	100	91	75

Example 3: Continuous Use of Lactase-RG Column — A column, identical to that described in Example 2, was prepared, and its activity measured, using a 3% lactose solution at pH 4.5 at a temperature of 45°C and a flow rate of 2 ml/min. The column was used for a period of 6 weeks, during which time the amount of lactose applied, the temperatures, and the flow rate were varied. During this period the ratio of the total amount of lactose hydrolyzed to the amount of enzyme on the column was in excess of 100,000 to 1. At the end of 6 weeks the activity was again measured as above. The results are summarized below.

Time (Weeks)	Hydrolysis (%)
0	64
6	64

Example 4: Preparation of Invertase-RG — The starting enzyme was Sucrovert (Numoline Division of Sucrest Corp), the invertase content of which was estimated to be about 0.5 to 1.0%. The resin was a commercial phenol-formaldehyde resin (Duolite S-30).

The resin was sieved to obtain particles of 20 to 30 mesh. After washing the resin with distilled water, it was soaked overnight in 0.1 M (0.58%) aqueous sodium chloride and then rewashed with distilled water. An aqueous solution of 50 ml of the invertase preparation was added to 20 grams of wet, drained resin at 25°C. The mixture was stirred gently for 15 minutes and then left at 5°C overnight, after which 2 ml of a 25% aqueous glutaraldehyde solution was added. The reaction mixture was left in the cold (5°C) for 16 hours and then rinsed thoroughly with distilled water. The invertase-RG exhibited 9% of the activity of the original enzyme.

Example 5: Invertase-RG Hydrolysis of Sucrose — Invertase-RG prepared as described in Example 4 was packed into a 1.2 cm x 15 cm jacketed column over a bed of sand. The column was washed with water. Aqueous sucrose solutions (10 and 30% in 0.02 N sodium acetate buffer, pH 4.5) were then fed through the column at varying flow rates. The column temperature was regulated at 37°C by passing warm water through the column jacket. The extent of hydrolysis was determined by analyzing for glucose in the effluent. The results are summarized below.

	Sucrose Concentration in Solution					
	- - - - - - - - - - 10% - - - - - - - - - - -				- - - 30% - - -	
Flow rate, ml/min	0.2	0.5	1.0	2.0	0.2	0.5
% hydrolysis	100	100	100	>95	100	90

POROUS GLASS AS A SUPPORT

Although enzymes are remarkably stable when adsorbed or coupled to the surface of inorganic carriers, the amounts of enzymes that can be insolubilized on the surface are rather low due to the limited surface area available. In addition, if enzymes are chemically coupled to inorganic carriers there are cumbersome and costly multistep procedures which must be followed.

Even when a porous inorganic carrier is used to insolubilize enzymes, the amount of enzyme that can be attached to the carrier is limited by the surface area that is available since only the enzymes in contact with the surface area will be retained for long periods of time. Attempts to more fully utilize the available volume in the pores of inorganic carriers have been unsuccessful. Attempts have been made to cross-link enzymes within the pores of inorganic carriers. However, such attempts have not been able to overcome the problem of moving the cross-linking agent from an outer solution into the pores without displacing the enzymes within the pores. Further, when cross-linking agents were introduced into the pores first, it was found impossible to later introduce the enzyme into the pores without an outflowing of the cross-linking agent. For example, a well-known water-soluble cross-linking agent such as glutaraldehyde could not be used to cross-link enzymes within the pores of inorganic carriers because of the above problems. Thus, there has been no known way to fully utilize the porous volume of inorganic carriers to immobilize enzymes.

R.A. Messing; U.S. Patent 3,804,719; April 16, 1974; assigned to Corning Glass Works; has shown that enzymes can be cross-linked within the pores of an inorganic carrier by choosing a water-insoluble cross-linking agent. The use of an essentially water-insoluble cross-linking agent, then, offers two paths for successfully accomplishing both adsorption and cross-linking within the pores and on the surface of the carrier.

Procedure 1: Enzymes are first adsorbed from an aqueous solution into the pores and on the surface of an inorganic carrier such as porous glass. The carrier is then removed from the aqueous enzyme solution. The glass is then exposed to or contacted with an organic solvent such as alcohol, ether, and the like, containing the cross-linking agent for sufficient time to assure cross-linking within the pores. Since the enzyme is essentially insoluble in the organic solvent it cannot move out of the pores into the outside solvent. However, the cross-linking agent will migrate into the pores and bring about cross-linking because it will remain in solution with varying concentrations of organic solvent and water. After cross-linking of the enzymes within the pores of the carrier, the carrier is removed from the organic solvent.

Procedure 2: Alternatively, an organic solvent such as alcohol, ether, or the like, containing the essentially water-insoluble cross-linking agent is first added to or contacted with the dry, porous carrier in either minimum volume such that the solution is completely absorbed by the porous carrier present, or in a larger volume with the subsequent removal of excess solvent by decantation or evaporation. Then, the porous carrier with water-insoluble cross-linking agent is exposed to an aqueous enzyme solution. The enzyme migrates into the pores where it is adsorbed and cross-linked, but the water-insoluble cross-linking agent cannot readily move out of the pores and into the aqueous medium. After the enzyme has been adsorbed and cross-linked within the pores of the carrier, the carrier is removed from the aqueous enzyme solution.

By means of the above procedures, three goals are accomplished. First, a greater amount of the pore volume is utilized for insolubilizing enzymes, thus permitting greater amounts of enzyme to be insolubilized by a given amount of carrier. Second, since a greater amount of enzyme is securely contained within the protective pores, less enzyme will be lost when the insolubilized enzyme composite is exposed to a turbulent reaction environment. Third, because of the greater rigidity of the inorganic carrier, a higher degree of enzyme immobilization is attained within the pores. The water-insoluble cross-linking agent used was 1,6-diisocyanatohexane, hereinafter referred to as DICH. The inorganic carrier used was porous glass.

It should be noted, however, that numerous carriers may be used as long as they have a porosity such that enzymes can be both adsorbed and cross-linked within the pores. The net effect is a greater utilization of pore volume and the provision of an environment for the enzyme which is less turbulent than that available when the pore volume was not so utilized. Thus, various porous inorganic bodies may be used, e.g., porous glass, dried inorganic gels such as those of Al_2O_3 and TiO_2, and naturally occurring porous inorganic bodies (e.g., diatomaceous earth), each having average pore diameters which will provide protection for the adsorbed and cross-linked enzymes and a generally large surface area for loading the enzyme. Preferably, the average pore diameter is between about 300 to 1000 A, depending on such factors as enzyme and substrate sizes, to assure maximum protection against the turbulence and desirably large surface area (protected by the pores) per gram of carrier.

Example 1 (Procedure 1): Papain solution: 200 mg of crude papain was diluted to 10 ml with 0.1 M phosphate buffer pH 8.1, dissolved, and placed in a 5°C water bath. 100 mg of porous glass was transferred to a 10 ml graduated cylinder and reacted in the 5°C bath with reciprocal shaking for 25 minutes. Shaking was then stopped and the adsorption was allowed to proceed at 5°C without shaking for about 14½ hours. Adsorption was then continued for a 30 minute period with reciprocal shaking at 5°C after which the remaining enzyme solution was decanted.

An ethyl ether solution containing 0.05 mg DICH per ml was precooled to 5°C after which 0.5 ml (0.025 mg DICH) was added to the glass in the 5°C bath. The sample was then reciprocally shaken with the cross-linking agent for 4½ hours at 5°C. The sample and cross-linking agent was then diluted with 10 ml of distilled water, and shaking in the 5°C bath was continued for 15 minutes.

The sample was then washed thoroughly with water and an activator solution of cysteine and ethylenediaminetetraacetic acid (EDTA) over a fritted glass funnel. The cysteine-EDTA solution activates the enzyme by reducing it and removing metal ion inhibitors, and such an activation technique is well known in the art. The thus-treated glass was finally transferred to a 25 ml flask in which it was stored and analyzed. The same 100 mg sample was assayed with casein at pH 5.8, 40°C repeatedly over the course of storage in distilled water at room temperature. The assay results over about a two-month storage period are given below. The first figures in each pair of columns refers to the storage time in days at which the assays were made. The second figure represents the activity found in mg crude papain per gram of glass.

Days	Activity
0	11.4
1	10.8
3	11.2
6	12.4
8	13.6
13	10.8
17	9.0
27	8.4
43	6.6
51	6.6
61	4.2

Example 2 (Procedure 2): Papain solution: 100 mg papain was diluted to 5 ml with 0.2 M phosphate buffer, pH 8.1, allowed to dissolve and then placed in a 5°C water bath. A DICH solution of 0.05 mg DICH per ml in ethyl ether was precooled to 5°C. Porous glass (100 mg) was transferred to a 10 ml graduated cylinder. A 0.5 ml (0.025 mg) portion of the DICH solution was added to the glass in the cylinder and then placed in a cold room (8°C) for 30 minutes. The cylinder and its contents were then placed in a 40°C bath with reciprocal shaking. Ether was evaporated from the contents of the cylinder in the 40°C bath until the sample was dry (5 minutes). The cylinder and the treated glass were then transferred to a 5°C bath.

The 0.5 ml (10 mg) of the papain solution was added to the treated glass and allowed to react in the 5°C bath with reciprocal shaking for 2 hours. The shaking was then stopped and the reaction was allowed to continue in the 5°C bath for an additional 45 minutes, after which 10 ml of distilled water was added and the sample was again shaken in the 5°C bath for 5 minutes. The enzyme and water was decanted and the sample was thoroughly washed over a fritted glass funnel with water and then treated with an activator solution of cysteine and EDTA. The sample was stored, handled, and assayed in the same manner as in Example 1. The assay results over a two-month period are shown below.

Days	Activity
0	8.2
1	8.8
2	8.8
4	9.2
7	9.6
9	8.2
14	10.2
18	7.8
24	8.4
44	5.6
52	5.6
62	1.0

Example 3 (Procedure 1): Papain solution: 100 mg papain was diluted to 5 ml with 0.1 M phosphate buffer, pH 7.9 allowed to dissolve and then placed in a 5°C water bath. A DICH solution of 0.2 mg DICH per ml in methanol was prepared and precooled to 5°C.

100 mg of porous glass was placed in a 10 ml graduated cylinder and the cylinder was placed in a 5°C water bath. Then, 2 ml (40 mg) papain solution was then added to the glass and allowed to react without shaking for 1 hour and 50 minutes in the 5°C bath. The cylinder and contents were then reciprocally shaken in the 5°C bath for a two-hour period after which the reaction was allowed to proceed without shaking for 15½ hours. Then, the cylinder and its contents were reciprocally shaken for 15 minutes in the 5°C bath after which the enzyme solution was finally decanted.

The glass was then washed with two 10 ml volumes, followed by one 4.5 ml volume of distilled water in the 5° bath with reciprocal shaking over 30 minute intervals for each wash. A 0.2 ml (0.04 mg) aliquot of the precooled DICH-methanol solution was added to the glass and reacted in the 5°C bath with reciprocal shaking for 3 hours and 5 minutes after which the sample and cross-linking agent was diluted to 10 ml with distilled water and the reaction was continued with reciprocal shaking for an additional 1½ hours in the 5°C bath. The sample was then thoroughly washed with water over a fritted glass funnel. The sample was then stored, handled, and assayed in the same manner as in Example 1 and the assay results are given below.

Days	Activity
0	40.8
0, 1 hour	36.4
3	36.0
5	36.0
10	33.2
14	26.6
24	24.0
41	20.4
49	13.4
59	6.0

Example 4 (Procedure 2): The papain and DICH solutions were the same as those used in Example 3 above. 100 mg of porous glass was transferred to a 10 ml graduated cylinder and placed in a 5°C water bath. Then, 0.2 ml (0.04 mg) DICH-methanol solution was added to the galss. The solution was completely absorbed in the pores of the glass and the glass particles appeared to be dry. The cylinder containing the glass was reciprocally shaken in the 5°C bath for 25 minutes. Then, 2 ml (40 mg) of the papain solution was added and reacted without shaking in the 5°C bath for 1 hour and 50 minutes.

The cylinder and contents were then reciprocally shaken in a 5°C bath for a two-hour period after which the reaction was allowed to proceed without shaking at 5°C for 15½ hours. The cylinder and its contents were then finally reciprocally shaken in the 5°C bath for 15 minutes and the enzyme solution was then decanted. The glass was then washed in the same manner as Example 3. The sample was likewise stored, handled, and assayed as in Example 1, and the assay results are given below:

Days	Activity
0	32.4
0, 1 hour	30.4
3	31.4
5	28.8
10	29.8
14	19.8
24	23.6
41	21.2
49	13.4
59	6.6

By comparing the results of Examples 1 and 2 with the results of Examples 3 and 4, it appears clear that the amount of enzyme retained by the glass over a two-month period is related to the amount of enzyme originally contacted with the glass. However, when lesser amounts of the enzyme were used, as in Examples 1 and 2, the storage stability of the enzyme remained more nearly constant. Thus, in those applications where it would be desirable to be reasonably certain of the enzyme activity over a prolonged period of time, a similar ratio of enzyme to glass would be used.

SILICA AND POLYAMINES AS SUPPORTS

R. Haynes and K.A. Walsh; U.S. Patent 3,796,634; March 12, 1974; assigned to U.S. Secretary of Health, Education and Welfare have provided a method for insolubilizing normally water-soluble biologically active proteins while retaining biological activity which comprises (a) adsorbing protein molecules in an aqueous medium as a monolayer on colloidal particles having a specific gravity greater than that of the protein and a net electric charge opposite to that of the protein at the pH at which adsorption and the subsequent cross-linking occurs; and (b) reacting the adsorbed protein molecules with a cross-linking agent having two or more groups reactive with side chains of amino acid residues to form stable covalent linkages between adjacent protein molecules.

In the product, insolubilized protein molecules envelop the colloidal adsorbent particle in a monolayer, and one object, i.e., that essentially all insolubilized protein molecules directly contact the surrounding medium is attained thereby. The insolubilized protein resists denaturation. The insolubilization of biologically active protein occurs with substantial retention of biological activity. The insoluble protein by reason of its size and hydrophilic nature can be readily dispersed in substrate-containing media, yet can be readily and quantitatively recovered by, e.g., centrifugation.

With reference to Figure 2.1a, the process is schematically depicted as proceeding sequentially by the adsorption of protein molecules indicated as relatively small, light-colored spheres onto colloidal adsorbant particles, followed by reaction with a multifunctional cross-linking agent to staple the adsorbed protein molecules about the adsorbent particle through formation of covalent linkages between adjacent molecules. As is apparent from the pictorial representation of Figure 2.1b, each protein molecule in the ultimate product is in direct contact with the surrounding medium (not shown).

The active sites of the biologically active protein molecules, shown as black dots are randomly oriented in the ultimate product. Nevertheless, many of them are presented to the exterior of the product and those distal to the outer surface have been shown to remain available to substrates penetrating the interstices between bound protein molecules. The covalent cross-linking between protein molecules appear to enhance resistance to denaturation by reducing the degrees of freedom enjoyed by individual molecules. For example, native trypsin is instantaneously deactivated in 8 M urea. Trypsin adsorbed on silica and cross-linked with glutaraldehyde has retained 50% activity after 30 minutes in 8M urea.

Any biologically active protein can be insolubilized. Preferably, the proteins have

more than two exposed amino groups for participation in the cross-linking reaction, i.e., more than two epsilon amino groups of lysyl residues in the protein
molecule. Of course, other groups borne by a particular protein molecule can
be employed in cross-linking, notably sulfhydryl groups of cysteine amino acid
residues and phenolic groups of tyrosine amino acid residues. Exemplary biologically active proteins which have been insolubilized include bovine trypsin,
bovine serum albumin, alpha-chymotrypsin, soybean trypsin inhibitor, hen's
egg white lysozyme, bovine pancreatic ribonuclease, ovalbumin, and rabbit
gamma G globulin.

FIGURE 2.1: SILICA AND POLYAMINES AS SUPPORTS

a.

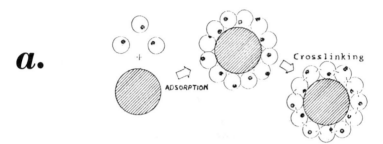

The Process Illustrated in Cross-Section

b.

A Product of the Process

Source: R. Haynes and K.A. Walsh; U.S. Patent 3,796,634; March 12, 1974

As exemplary of the wide range of proteins believed suitable for insolubilization
can be mentioned amylase, cellulase, dehydrogenase, depolymerase, glucoamylase,
isomerase, lipases, pectinases, penicillin amidase, proteases such as bromelain and
papain, pullulanase, pheylalanine hydroxylase, homogentisic acid oxidase, L-as-
paraginase, and tyrosinase.

The colloidal particles onto which the biologically active proteins are adsorbed
are chosen to exhibit specific gravity greater than that of the protein to be adsorbed so that protein recovery by centrifugation or the like is facilitated. Generally, then, the colloidal adsorbent exhibits specific gravity greater than about
1.3. The specific gravity of the preferred adsorbent, colloidal silica, is on the
order of about 2. The adsorbent particles employed are of colloidal dimensions

such that the insolubilized protein can be readily dispersed in substrate-containing media, yet the protein packages can be easily recovered by centrifugation or the like by reason of their cores of relatively high specific gravity adsorbent. Generally, the adsorbent particles range in size from 50 to about 20,000 A, preferably from about 100 to 200 A, and most preferably are about 150 A in diameter.

The particles are preferably chosen, in terms of size, from the smaller end to the foregoing broad range in order to maximize surface area per unit weight. In addition to colloidal silica, the preferred adsorbent, there can be mentioned as candidates for adsorbent use activated charcoal, hydroxyapatite, alumina C gamma, and bentonite. Preferably, as in the case of colloidal silica, the adsorbent particles employed are essentially nonreticulate so that access to bound protein is not restricted by the necessity that the substrate molecules penetrate the adsorbent particle.

For the sake of protein adsorption, it is necessary that at the pH at which adsorption and cross-linking occur the colloidal adsorbent bear a net electric charge opposite that of the protein molecules so that ionic bonding aids other noncovalent bonding forces such as, e.g., hydrogen bonds. In particular instances the optimal pH for the cross-linking reaction may differ substantially from that at which adsorption is best effected on the adsorbent particle per se. Generally, protein molecules are negatively charged at pH greater than their isoelectric point (the pH at which the molecule is electrically neutral). Conversely, they are generally positively charged below their isoelectric point. Until cross-linking occurs to staple the protein envelope about the colloidal adsorbent, the protein molecules are but weakly held by noncovalent bonds.

Consequently, if the isoelectric point is interposed between optimal pH for cross-linking and optimal pH for adsorption onto the adsorbent per se, crossing the isoelectric point to cross-link can result in desorption of protein molecules by charge repulsion. As an example, it has been found that adsorption of ovalbumin on negatively charged colloidal silica drops off rapidly above pH of about 5. If it is desired to cross-link ovalbumin on silica with glutaraldehyde, the preferred cross-linking agent, the reaction must proceed at pH greater than about 5.5 and the ovalbumin desorbs at that pH.

Difficulties like the foregoing are obviated by appropriate surface treatment of the adsorbent particle per se, from whence springs the general applicability to all proteins regardless of isoelectric point. In the case of silica, problems arising in a particular instance are eliminated by adsorbing a surface coat of, e.g. polyethylenimine onto the silica surface prior to adsorption of protein. The polyamine coat is then cross-linked by reaction with conventional amine-reactive cross-linking agents. Subsequently, the negatively charged protein "sees" a positive surface on the adsorbent particle and can be readily adsorbed at a pH optimal for the protein-cross-linking reaction.

Any macromolecular polyamine subject to cross-linking can be employed to alter the surface charge of the particle per se, e.g., polyethylenimine, polylysine, polyornithine, etc. In the light of the foregoing, of course, other manners of altering the net charge of adsorbent particles will occur. In any case, the colloidal particles hereinafter referred to encompass both colloidal substances per se like colloidal silica, and colloidal particles of, e.g., silica bearing a polyamine surface coat or otherwise altered in charge properties to optimize protein retention during cross-linking.

Colloidal silica is employed as the preferred adsorbent (Ludox HS, diameter about 150 A, SG about 2). In order to avoid competitive formation of soluble silicate complexes with protein, the silica was routinely dialyzed against water prior to use. Sodium borate is employed to buffer the various solutions. Sodium borohydride is added to reduce excess cross-linking agent so that in the following inhibition demonstrations no cross-linking occurs between protein and inhibitor and, in general, further reaction of available cross-linking groups is effectively precluded. The amount of cross-linking agent to provide maximum product stability with minimum inactivation can be readily determined by the art-skilled. Generally an excess is employed and in the case of trypsin a fivefold excess of glutaraldehyde has been used.

In commercial operation, excess cross-linking agent can be removed by water wash or other conventional means. Desirably, sodium chloride or other electrolytic salts are added where indicated to aid in sedimentation of the bound protein. Thereby, the net charge of bound protein is altered so that charge repulsion is neutralized and aggregation promoted. Subsequently, the bound protein can be redispersed and dialyzed by conventional techniques to to remove salt.

Example: This example demonstrates the very high retention of activity toward normal substrates enjoyed by the insolubilized protein by comparison to the native water-soluble protein; and also establishes the retention of activity of active sites distal to the exterior of the bound protein by comparison with retained activity toward casein, a large substrate.

100 mg lots of trypsin are rendered insoluble by treatment with a stoichiometric amount of silica (silica: trypsin 2.3:1 w/w, 15 minutes, room temperature) and 125 μl of 25% glutaraldehyde for 1 hour at room temperature. Both reactions were carried out in 50 ml of 0.1 M borate, 0.001 M benzamidine, pH 8.5. The product, in this case, is then further treated with 0.05 M $NaBH_4$ for 20 minutes at 0°C to reduce the remaining aldehydic groups and thus prevent covalent cross-linking to the protein inhibitors used in subsequent experiments. The recovery of insoluble trypsin after the glutaraldehyde reaction is greater than 99%, as judged by activity towards BAEE (benzoyl arginine ethyl ester). However, losses in subsequent transfers reduce the overall yield to 80%. Enzymatic properties of a typical preparation are given in the following table.

	Mols Active Site */Mol Protein	BAEE Activity** Molecules/Active Site/Minute	Relative Caseinolytic Activity*** per Active Site
Soluble trypsin	0.74	1,700	100
Insoluble trypsin	0.60	1,500	17.5

*Active site titrations by the method of Chase, T. Jr., et al, *Biochem. Biophys. Res. Commun.* 29, 508 (1967), using p-nitrophenylguanidobenzoate. The insoluble enzyme was removed by centrifugation just prior to reading the absorbance.

**0.01 M BAEE in 0.01 M tris amine buffer, pH 7.8, 26°C in a pH stat.

***Method of Laskowski, Sr., *Methods in Enzymology* 2, 26 (1955).

It should particularly be noted that BAEE turnover in the insolubilized trypsin is fully 88% of that obtaining in the native enzyme, so that the advantages of insolubilization are attained without undue loss of activity. The sharply decreased turnover of the large casein substrate when compared to the substantially retained turnover toward the smaller BAEE substrate is persuasive that active sites distal to the exterior of the protein package can yet be utilized by substrates which can penetrate the interstices between bound protein molecules.

CROSS-LINKING WITHIN BACTERIAL CELLS

Glucose isomerase is produced by fermentation of organisms, such as *Pseudomonas hydrophila, Streptomyces flavovireus, Streptomyces echinatur, Streptomyces achromogenus, Streptomyces albus, Streptomyces olivaceus* and the like, in appropriate nutrient media. The glucose isomerase is formed inside the bacterial cells which grow during its production. The cells can be filtered off from the fermentation beer and used directly as a source of glucose isomerase. Alternatively, the cells can be recovered by filtration and then ruptured by well-known means. The resulting ruptured cells and released contents can be used as a source of glucose isomerase. Still further, the cells can be ruptured and the debris removed by centrifugation. The supernatant liquid can be used directly as a source of glucose isomerase or the enzyme can be recovered as a powder from this liquid by well-known techniques.

The glucose isomerase is most suitable for recovery and reuse if it is still contained within the original bacterial cells. The bacterial cells can be easily separated from the sugar-containing reaction conversion medium. However, the prior techniques still prohibited significant reuse of the bacterial cells since the cells lost about 50% of their glucose isomerase activity during each use.

M.F. Zienty; U.S. Patent 3,779,869; December 18, 1973; assigned to Miles Laboratories, Inc. has developed a process for the stabilization of glucose isomerase in bacterial cells which comprises treating bacterial cells containing glucose isomerase activity with glutaraldehyde.

The bacterial cells containing glucose isomerase activity can be produced by well-known procedures. The preferred enzyme-containing cells are produced by growing under submerged aerobic conditions a culture of *Streptomyces olivaceus* NRRL 3583 or mutants thereof in a medium containing appropriate nutrients. The resulting bacterial cells are separated from the fermentation beer by filtration or centrifugation. The recovered bacterial cells are then suspended in an aqueous medium and are mixed with glutaraldehyde. The glutaraldehyde is employed in an amount from about 0.1 to about 50 weight percent based on the dry weight of the cells. Preferably, the glutaraldehyde is employed in an amount from about 10 to about 50 weight percent based on the dry weight of the cells.

The bacterial cells initially have a pH of about 8.5. As the bacterial cells are treated with the glutaraldehyde, the pH drops eventually to about 6.5. If the initial pH is above about 8.5 or if the pH during treatment drops to below about 6.5, suitable buffering materials should be added to maintain the pH of the bacterial cells within the range from about 6.5 to about 8.5. The bacterial cells should be treated with the glutaraldehyde for from about ½ to about 2 hours. The preferred treatment time is from about 1 to about 1½ hours. The treatment temperature is not narrowly critical and can conveniently be from about 15°C to about 60°C.

Example 1: A culture of *Streptomyces olivaceus* NRRL 3583 was transferred to an agitated aerated fermentor containing 10 liters of an aqueous mixture containing 0.7% xylose, 0.3% refined corn starch, 1.0% peptone, 0.50% meat extract, 0.25% meat extract, 0.25% yeast extract, 0.50% sodium chloride and 0.05% magnesium sulfate and having a pH of 7.0. All the above percent values were on a weight/volume basis.

The agitator was rotated at 400 rpm and air was passed through the medium at a rate of 3 volumes of air per volume of medium per minute. The fermentation was continued for 24 hours at 32°C. The fermentation beer was then centrifuged at 40,000 rpm to separate the bacterial cells. A portion of the bacterial cells was then frozen and lyophilized.

Three grams of the freeze-dried whole cells were suspended in 50 ml of aqueous tris(hydroxymethyl)aminomethane at pH 8.5, and the stirred suspension was treated with 0.1 gram of a 25 weight percent aqueous glutaraldehyde solution (0.83 weight percent glutaraldehyde based on dry weight of the cells) with stirring at room temperature (about 20°C) for 1.5 hours.

The treated cells having an activity of 145 GIU/g were collected by filtration and added to a 62.5 weight percent aqueous solution of glucose containing 0.001 weight percent cobalt chloride and 0.01 weight percent magnesium chloride. The resulting mixture was reacted at 70°C for 2 hours to convert about 4.5% of the glucose to fructose. The cells were separated from the reaction mixture by centrifugation and the fructose-dextrose mixture was decanted.

The cells were assayed to have the same glucose isomerase activity as before their use. The cells were then used again to treat a fresh glucose solution under the same conditions as described above and were recovered and assayed. This procedure was repeated for a total of twelve uses of the cells. These cells still contained 56% of the initial glucose isomerase activity. Cells which had not been treated with glutaraldehyde lost substantially all their glucose isomerase activity upon being used twice.

Example 2: A fermentation beer containing *Streptomyces olivaceus* NRRL 3583 bacterial cells was prepared in the manner described in Example 1. A portion of this beer was centrifuged and a 3 gram portion of wet whole bacterial cells was isolated. These cells were then suspended in 50 ml of water and treated with 0.1 gram of a 25 weight percent aqueous glutaraldehyde solution in the pH range of 6.5 to 8.5 at room temperature for 1 hour. The treated cells having an activity of 265 GIU/g were collected by filtration and were used to convert glucose to fructose in the manner described in Example 1. The cells were recovered and reused with fresh glucose for a total of 12 times. The cells still retained 60.7% of the initial glucose isomerase activity.

Example 3: A fermentation beer containing *Streptomyces olivaceus* NRRL 3583 bacterial cells was prepared in the manner described in Example 1 employing a 100 gal fermentor. This beer had a total activity of 1,350,000 GIU and contained 7 lb of bacterial cells on a dry weight basis (425 GIU/g). The pH of the beer was adjusted to 8.2 by addition of sodium hydroxide. A 7 lb portion of 50 weight percent aqueous glutaraldehyde solution diluted to a concentration of 0.5 to 0.6% (weight/volume basis) was added to the beer with agitation over a 30 to 40 min period (50 weight percent glutaraldehyde based on dry weight of cells).

The resulting mixture was then reacted at room temperature for 1.5 hours with mild agitation. A 0.5% (weight/volume basis) portion of diatomaceous earth filter aid was then added and the beer was filtered on a rotary vacuum filter. The filter cake was washed with water to remove spent beer residues. The collected bacterial cells had a total activity of 1,215,000 GIU or 90% of the initial activity.

A glucose solution was prepared by mixing 225 lb of glucose in 55 gal of water (49.2% glucose on a weight/volume basis). To this solution were added cobalt chloride in an amount of 0.001% (weight/volume basis) and magnesium chloride in an amount of 0.01% (weight/volume basis). A portion of the above-prepared glutaraldehyde-treated bacterial cells having a total activity of 407,250 GIU was also added to the glucose solution. The resulting mixture was stirred at 60°C for 24 hours during which time the pH was maintained at 7.7 to 7.9 by addition of sodium hydroxide.

The bacterial cells were filtered off from the reaction mixture which was then assayed to contain 40.2% fructose. The above-collected bacterial cells were then reused on a fresh glucose solution in the manner described above. This procedure was then repeated until the bacterial cells had been used a total of 9 times. In the ninth run the glucose isomerase activity in the bacterial cells produced 38.8% fructose, which is still 96.5% of the initial performance.

ALKYLENE DIAMIDINE LINKAGE TO POLYAMINE CELLULOSE

According to *L.E. Weeks and J.H. Reynolds; U.S. Patent 3,741,871; June 26, 1973; assigned to Monsanto Company* an enzymatically-active, water-insoluble composite comprises fibrous cellulose, a polyamine covalently attached to the cellulose by a carbamic linkage, and an active enzyme cross-linked to the polyamine by an alkylene diamidine linkage. The composite is useful as a packing for enzyme reactors.

The composite is prepared by reacting in an aqueous medium fibrous cellulose, with a cyanogen halide such as BrCN or ClCN so as to produce a pendant iminocarbonic acid ester group on the cellulose, reacting the iminocarbonic acid ester group with a polyamine to covalently attach polyamine to cellulose by means of a carbamic linkage, and thereafter coupling a free primary amino group of the attached polyamine and an amino group of an active enzyme which is not essential for enzymatic activity by means of a lower alkyl diester of saturated aliphatic diimidic acid to form an alkylene diamidine linkage between enzyme and polyamine. A suitable matrix or carrier is fibrous cellulose or insoluble derivatives such as methylcellulose, carboxymethylcellulose, viscose, and the like. Hydroxyl groups present on the fibrous cellulose react with the cyanogen halide to form a pendant iminocarbonic acid ester group substantially according to the reaction

$$\begin{array}{c}\diagdown\text{CH--OH}\\|\\\diagup\text{CH--OH}\end{array}\quad\xrightarrow{\text{XCN}}\quad\begin{array}{c}\diagdown\text{CH--O--C}{\equiv}\text{N}\\|\\\diagup\text{CH--O--H}\end{array}\quad\longrightarrow\quad\begin{array}{c}-\text{CH--O}\diagdown\\|\qquad\qquad\text{C=NH}\\-\text{CH--O}\diagup\end{array}$$

wherein X designates a halogen atom such as chlorine or bromine. Thereafter the treated cellulose, bearing iminocarbonic acid ester groups, is treated with a polyamine so as to react a primary amino group of the polyamine with the iminocarbonic acid ester group of the cellulose and to form a carbamic linkage between the polyamine and the fibrous cellulose.

The polyamine moiety acts as a spacer for the active enzyme and holds the enzyme sufficiently far away from the cellulose matrix to minimize steric inhibition of enzyme activity. Suitable polyamines for this purpose are the aliphatic diamines represented by the formula

$$\text{H}_2\text{N}-\left(\text{CH}_2\right)_n-\text{NH}_2$$

where n is an integer having a value of 3 to 18, inclusive, alkylene dianilines containing 1 to 18 carbon atoms in the alkylene portion thereof, and polyalkylene imines containing primary, secondary and tertiary amino groups and having a molecular weight in the range of about 100 to about 50,000.

The cellulose matrix having pendant primary amino groups connected to the polysaccharide chain of the cellulose by a relatively long hydrocarbon chain covalently bonded thereto via a carbamic linkage is then coupled to an active enzyme using a lower alkyl diester of saturated aliphatic diimidic acid so as to produce an alkylene diamidine linkage between the enzyme and the polyamine. The diimidic acid can contain 1 to 8 carbon atoms in the alkylene portion thereof. Illustrative cross-linking agents are the lower alkyl diesters of malonimidic acid, succinimidic acid, glutarimidic acid, adipimidic acid, pimelimidic acid, suberimidic acid, and the like.

Example 1: Activation of Cotton with Cyanogen Bromide — Cyanogen bromide (50 g) and cold water (2.5 l) were combined in a four-liter beaker and placed in an ice bath on a magnetic stirrer plate. pH of the solution was immediately brought in the range of 11 to 12.5 by addition of an aqueous 2 N sodium hydroxide solution. Next, chopped clean cotton roving (50 grams, 1 to 2 inches long) was added to the solution and the solution was manually stirred and maintained at a pH of about 11 to 12.5 by the addition of aqueous 2 N sodium hydroxide solution or 2.5 N sodium hydroxide solution. Stirring and the pH were maintained for about one hour at which time the pH stabilized at 12.0. At that time 200 ml of 2 N sodium hydroxide solution and 153 ml of 2.5 N sodium hydroxide solution had been added.

Thereafter the resulting BrCN-activated cotton was washed 8 times with two-liter

aliquots of cold water and 4 times with two-liter aliquots of 0.1 M aqueous sodium bicarbonate solution. Final pH was observed to be 8.4. During each wash
cycle the activated cotton was stirred and after each wash cycle the activated
cotton was filtered.

Example 2: Attachment of Hexamethylene Diamine — Cold activated cotton
prepared in Example 1, above, was added to hexamethylene diamine (50 grams)
in cold 0.1 M aqueous sodium bicarbonate solution (2 liters). The pH of the
solution was adjusted to 8.6 by the addition of 4 N aqueous hydrochloric acid
solution before the addition of activated cotton took place. The resulting mixture was stirred in an ice bath for one hour. pH of the mixture was observed
to drop rapidly to 8.05.

Thereafter the hexamethylene diamine-treated cotton was washed with stirring
and filtered after each wash cycle, followed by the passing of an equal volume
of water of solution through the filter. The following wash sequence was used:

(a) water—two liters each time, twice cotton was stirred with two liters
of water and filtered, followed by two liters of additional water
passed through the filter after each filtration.

(b) 0.05 N HCl—two liters, stirred and filtered, pH of filtrate 2.0.

(c) water—same procedure as in (a).

(d) 0.1 M NaHCO₃—same procedure as in (a).

(e) 0.5 M NaCl—same procedure as in (a).

(f) water—two liters each time, thrice stirred with two liters of
water and filtered, followed by two liters of water passed
through the filter after each filtration.

The hexamethylene diamine-treated cotton was then stored at about 0° to 5°C
for about three days and tested for the presence of amino groups using Crocein
Scarlet MOO solution. The test gave substantive red color to the cotton, i.e.,
the test was positive for amino groups. Untreated cotton subjected to the same
test procedure was not dyed. Acetone washed hexamethylene diamine-treated
cotton also gave a positive indication for the presence of amino groups.

A sample of hexamethylene diamine-treated cotton was then washed with acetone
and dried at 50°C in a vacuum oven. Test for the presence of amino groups on
the dried sample using Crocein Scarlet MOO solution was positive. Thereafter all
of the hexamethylene diamine-treated cotton prepared as set forth hereinabove
was washed 5 times in two liters of acetone with stirring and filtering between
wash cycles. After the acetone was removed the cotton was dried in vacuum at
50°C for four hours. Yield: 50.9 grams.

A sample of the hexamethylene diamine-treated cotton was analyzed for nitrogen.
An average nitrogen content of about 1.09% by weight was found. The analytical
results indicate that about 12 to 13% of the cotton glucoside units have hexamethylene diamine attached thereto.

Example 3: Binding of Papain to Hexamethylene Diamine-Treated Cotton —
Hexamethylene diamine-treated cotton was packed into a stainless steel nipple

(1" i.d. x 3") with stainless steel caps and plastic porous discs as fiber retainers fitted at each end of the nipple. The caps were provided with plastic tubing fittings.

Dimethyladipimidate (2.00 g) was dissolved in water (150 ml). Aqueous 1 N sodium hydroxide solution was then added thereto to rapidly bring pH to 10. The resulting solution was then recirculated through the packed hexamethylene diamine-treated cotton at pH 10 for five minutes and at a rate of 132 ml/min. pH was lowered quickly to 6.5 during recirculation in order to slow the reaction. Thereafter, the packed cotton was quickly washed twice by recirculating for five minutes two 2-liter aliquots of water.

Purified papain (2.0 g) was dissolved in water (200 ml) and the pH adjusted to 10 by 1 N aqueous sodium hydroxide solution. This solution was already prepared. The obtained solution was then recirculated without delay through the packed, dimethyladipimidate-treated cotton at pH 10 for 30 minutes at a rate of 132 ml/min.

Thereafter the packed cotton was washed 3 times with 2-liter aliquots of water by recirculation for five minutes each at a rate of 132 ml/min. Material balance on the papain remaining in the recirculating solution and the first wash aliquot (2nd and 3rd wash aliquots contained no papain) indicated that about 26 mg of papain had been attached per gram of the cotton.

Example 4: Cross-Linking of Papain and Polyethyleneimine-Treated Cotton —
A reactor was prepared by packing polyethyleneimine-treated cotton (10 g) into a stainless steel nipple (1" i.d. x 3") with stainless steel caps provided with plastic tubing fittings and plastic porous discs as fiber retainers fitted at each end of the nipple. Purified papain (2.0 g) was dissolved in water (100 ml), pH of the solution was adjusted to 6.1 by the addition of dilute aqueous sodium hydroxide solution, and the solution then circulated through the packed reactor at a rate of 132 ml/min. During initial circulation pH of the solution was observed to rise from 6.1 to 7.1.

Dimethyladipimidate (2.0 g) was then added dry to the circulating solution over a five-minute interval while maintaining the solution at a pH of 10 by addition of 1 N aqueous sodium hydroxide solution. After dimethyl-adipimidate addition was completed, the solution was recirculated for an additional 50 minutes and pH was maintained at 10 by the addition of 3 N aqueous hydrochloric acid solution. Thereafter the packed cotton was washed three times with 2-liter aliquots of water by recirculation for five minutes each at a rate of 132 ml/min. The spent recirculating solution and the wash solutions were then checked for the presence of papain by measuring absorbance of the solutions at 280 nm. The obtained results are presented in the table below.

Solution	Volume, ml.	A_{280} at dilution		pH
Spent recirculated solution.	65	0. 928	1:100	9. 4
Do	65	0. 390	1:250	
First wash	1,980	0. 666	1:5	
Second wash	1,920	0. 154	1:1	
Third wash	1,900	0. 027	1:1	

Enzymatic activity was evaluated using a benzoyl-L-arginine ethyl ester solution. The solution was pumped through the reactor at a steady rate and hydrolysis of the substrate was monitored by measuring absorbance at 253 nm in a 0.1 mm cell. The experimental data are compiled in the following table.

Flow rate, ml./minutes	Time from start, minutes	Percent hydroly- sis of substrate
12	30	56
12	45	60
12	60	60
12	70	60
6	70–270	[1] 85

[1] Steady

Hydrolysis rates were steady and no wash-off of the papain was observed.

GLUTARALDEHYDE CROSS-LINKED ENZYME-TANNIC ACID COMPLEX

In the procedures of *W.L. Stanley and A.C. Olson; U.S. Patent 3,736,231; May 29, 1973; assigned to U.S. Secretary of Agriculture* insolubilized but active enzymes are prepared by reacting an enzyme with tannic acid and glutaraldehyde. The enzyme to be insolubilized is mixed with water to form a dispersion. (The term dispersion is used herein in a generic sense to include true solutions, colloidal solutions, or suspensions.) Where necessary, the pH of the dispersion is preferably adjusted to a level at which the enzyme is soluble.

One then adds thereto an aqueous solution of tannic acid. For best results this solution is first adjusted (e.g., by the addition of NaOH, KOH, or phosphate buffer) to a pH at which the enzyme in question is soluble, this pH level generally coinciding with that at which the enzyme is active. The amount of tannic acid is not critical. Usually a large excess, for example, 10 to 15 parts thereof per part of enzyme, is used and the unreacted residue removed in a subsequent washing step.

The enzyme-tannic acid mixture after stirring for a brief period, e.g., about 5 minutes, may be applied directly to the next step (treatment with glutaraldehyde). Alternatively, the enzyme-tannic acid complex may be removed from the mixture for use in the next step. In such event, it is preferred to hold the mixture for a substantial period of time, e.g., about 12 to 24 hours, to ensure complete precipitation.

Having prepared the enzyme-tannic acid mixture (or isolated complex), an aqueous solution of glutaraldehyde (usually in a concentration of about 25%) is added. The amount of glutaraldehyde is not critical. Usually, a large excess, for example, 10 to 15 parts thereof per part of enzyme, is used; the unreacted residue is removed in a subsequent washing step.

The resulting mixture (enzyme, tannic acid, and glutaraldehyde) is held for a period of time, preferably with mild agitation, to ensure formation of the enzyme-TAG product. Usually, the holding is for a period of about 12 to 24 hours in a cold room at about 1° to 10°C. The product is then collected by centrif-

ugation preferably with the use of a refrigerated centrifuge. To wash out unbound enzyme and reagents, the collected product is resuspended in fresh water
and recentrifuged. These steps of suspension in water and centrifugation are
preferably repeated until a product of desired purity is achieved. The so-prepared enzyme-TAG product is dried if it is to be preserved for future use. Drying from the frozen state (lyophilization) is preferred to avoid heat-damage to
the product.

For use in conducting enzymic reactions, the enzyme-TAG product can be used
directly, particularly if it exhibits a granular or crystalline form. Usually, it is
preferred to mix the complex with an inert carrier or support, preferably a solid
water-insoluble material in subdivided form. Since incorporation with a carrier
provides a mass through which water and other liquids can percolate readily,
the carrier can be incorporated with the crude product before the washing step,
whereby removal of unbound enzyme and reagents is facilitated. In a typical
practice the crude product is mixed with about 5 to 10 times its weight of a
carrier, and a column is packed with the resulting composition. Water is then
passed through the column in sufficient quantity to remove all water-soluble
components.

Example 1: Preparation of Lactase-TAG — The starting material was a commercial product, the lactase content of which was estimated to be about 10%. The
lactase preparation (0.3 g) was dissolved in 15 ml of doubly-distilled water (pH
7.0). To this was added 10 ml of 10% aqueous tannic acid solution which had
been adjusted to pH 4.5 with sodium hydroxide. After stirring for 5 minutes,
2.4 ml of 25% aqueous glutaraldehyde solution was added. The mixture was
stirred gently overnight at 34°F. The mixture was centrifuged for 10 minutes at
10,000 rpm in a refrigerated centrifuge. The liquid was discarded and the crude
product was suspended in distilled water and centrifuged. This purification process (suspending in water and centrifuging) was repeated 4 times and, finally, the
product was collected. The lactase-TAG complex exhibited 60% of the activity
of the original enzyme.

Example 2: Lactase-TAG Hydrolysis of Lactose — Lactase-TAG prepared as
described in Example 1 was added to 500 mg of Celite, a diatomaceous filter
aid. The mixture was thoroughly blended with a stirring rod to give a free-
flowing powder which was packed into an 8 mm jacketed column (2 cm long)
over a bed of sand, using gentle tamping with a stirring rod. The column was
then washed several times with pH 4.5 (0.02 M) phosphate buffer.

A lactose solution (3.0% in 0.02 M phosphate buffer) was then fed through the
column at varying flow rates using a pressure of inert gas to aid the movement
of liquid from the feed reservoir. The column temperature was regulated by
passing warm water through the column jacket. The extent of hydrolysis was
determined by analyzing for glucose in the effluent. The results are summarized
below.

Temp., °C	36	37	37	36	45	45	55	55
Flow, ml./min.	0.36	0.41	0.58	0.71	0.25	0.61	0.42	0.62
Percent hydrolysis	86	90	83	80	96	93	96	96

Example 3: Glucose Oxidase-TAG — Glucose oxidase-TAG was prepared by a

method similar to that described in Example 1 from a commercial glucose oxidase preparation, estimated to contain 10% of active enzyme. In this case, 2 grams of the enzyme preparation, 2 ml of 10% aqueous tannic acid solution at pH 5, and 0.6 ml of 25% aqueous glutaraldehyde were used. A column was prepared as described in Example 2 and applied to a 0.35% glucose solution containing 0.3% hydrogen peroxide. The results are summarized below.

Temperature, °C	35	45
Flow, ml/min	0.33	0.33
Conversion, %	70	74

Example 5: Preparation of Catalase-TAG — A commercial catalase preparation (20 mg) was dissolved in 1 ml of water and to this was added 1 ml of 1% aqueous tannic acid solution which had been adjusted to pH 7.0. After stirring for 5 minutes, 0.3 ml of 25% aqueous glutaraldehyde solution was added, and the mixture stirred gently overnight at 34°F.

The mixture was then centrifuged for 10 minutes at 10,000 rpm in a refrigerated centrifuge. The liquid was discarded and the crude product was suspended in distilled water and centrifuged. This purification process (suspending in water and centrifuging) was prepared 4 times, and the product was then collected. The catalase-TAG product exhibited 52% of the activity of the original enzyme.

PHYSICAL TECHNIQUES

EXCHANGE ONTO RESIN SUPPORTS

Chelating Polymers

The solid phase enzymes prepared or proposed fall into four distinct categories as follows: (1) adsorption type, (2) ionic type, (3) entrapped type, and (4) covalent type. In the first type of solid phase enzyme the enzyme is adsorbed onto an inert support, such as glass beads, charcoal, or polysaccharides, for example cellulose. A major disadvantage of this type of solid phase enzyme is that because of the weakness of the physical adsorption bonding desorption of the enzyme may occur with changes in inter alia ionic strength, temperature and pH values.

In the second type of solid phase enzyme, the enzyme is bound by ionic bonding to a polyionic carrier, such as a methacrylic acid copolymer or another charged polymer in the form of an ion exchange resin. Unfortunately, the polyionic carrier rarely confers the desired level of stability on the attached enzyme.

In the third type of solid phase enzyme, the enzyme is physically entrapped within a polymeric matrix in the form of a molecular sieve such as a cross-linked polyacrylamide gel. The principal disadvantages of this type of solid phase enzyme is its relative inaccessability to large molecules and diffusion of the enzyme from the carrier with small amounts of the enzyme still leaking out even after exhaustive washing of the solid phase enzyme.

In the fourth type of solid phase enzyme, the enzyme is covalently bound under mild conditions to a reactive group of a polymer, such as acrylamide copolymers available under the trademark Enzacryls. Most modern work concerning solid phase enzymes has been concerned with this type of covalently bound material. However, the reactive conditions required for the covalent bonding have to be carefully controlled in order to avoid destruction of the

133

enzyme molecule and it is usually difficult to strip exhausted enzyme from the polymer to enable the polymer to be recharged with fresh enzyme.

S.A. Barker, J.F. Kennedy, and J. Epton; U.S. Patent 3,794,563; February 26, 1974; assigned to Aspro-Nicholas Limited, England have found that if polymers having chelating sites defined by pairs of adjacent hydroxy and carboxylic acid groups are used as solid supports for enzymes, the bond between the enzyme and support is such that many disadvantages of solid phase enzymes are overcome. In particular the enzyme can be attached to and stripped from the polymer under relatively mild conditions. Generally a desirable level of enzyme stability is shown by the solid phase enzyme.

According to the process, there is provided a complex of a biologically active protein such as an enzyme, antibody or antigen with a polymer having recurring chelating sites defined by pairs of adjacent hydroxy and carboxylic acid radicals, some or all of which sites optionally are chelated with metallic or the like ions. It is thought that the enzyme component is attached to the polymer at the chelating sites but the precise nature of the bonding has not yet been ascertained.

The enzyme-polymer complex has the enzyme bound to the polymer with sufficient strength for the polymer to constitute a solid support for the enzyme. Sometimes this bond strength is low enough to enable the enzyme to be readily stripped from the polymer by treatment with aqueous strong salt or aqueous buffer. However, sometimes the enzyme solubilizes in the presence of its substrate.

Polymers having chelating sites have been known for a good number of years and have been employed for such purposes as extracting metallic ions from aqueous solution. For the purposes of this process, the chelating sites on the polymer are defined by pairs of adjacent hydroxy and carboxylic acid radicals. Preferably the polymer has recurring aromatic nuclei on adjacent ring carbon atoms of which the pair of radicals are substituents.

More preferably, the polymers contain recurring ortho-hydroxybenzoic acid groups, the hydroxy and carboxylic acid radicals of which groups define chelating sites. Such polymers may be homopolymers of ortho-hydroxybenzoic acids having as a ring substituent an ethylenically unsaturated radical or copolymers of such an ortho-hydroxybenzoic acid derivative and one or more ethylenically unsaturated comonomers. The unsaturated radical may be of the formula $-X (CO)_n CR=CR_1R_2$ where each R independently represents hydrogen, alkyl of one to six carbon atoms or halogen, n represents 0 or one and X represents a direct bond, oxygen, sulfur, $-NH-$ or methylene.

Preferred unsaturated radicals are those of the formula $-X (CO)_n CR=CH_2$ where R represents methyl or, more preferably, hydrogen, n and X are as defined supra but more preferably n is one and X is $-NH-$. Advantageously, the ethylenically unsaturated radical is in the four or five position of the phenyl ring.

Examples of the preferred polymers are homopolymers and copolymers, especially with acrylamide, of N-acryloylaminosalicylic acids, for example N-acryl-oyl-4- or 5-aminosalicylic acids. Such polymers may be cross-linked with, for example, N-methylene-bis-acrylamide.

Some or all of the chelating sites of the polymer may be chelated with metal-lic or the like ions, for example titanium or borate ions. The enzymes which can be attached to any particular polymer will be dependent upon the precise nature of the enzyme and of the polymer. Such factors as steric hindrance may, for example, prevent a particular enzyme from being bound to a partic-ular polymer. However, the suitability of a particular polymer as a solid sup-port for a particular enzyme may be determined by simple experimentation. Examples of specific enzyme/polymer systems which have been shown to be satisfactory as solid phase enzymes are as follows:

β-glucosidase/titanium complex of N-acryloyl-4-aminosalicylic acid homopolymer;

lactate dehydrogenase/N-acryloyl-4-aminosalicylic acid homopolymer;

α-amylase/titanium complex of N-acryloyl-4-aminosalicylic acid homopolymer;

glucamylase/titanium complex of N-acryloyl-5-aminosalicylic acid homopolymer;

The polymer-enzyme complexes may be formed by simply mixing the enzyme with aqueous suspensions of the polymer and subsequently separating the solids content by, for example, a centrifuge, from the supernatant liquor.

The solid phase enzymes may be used in place of the soluble and often rela-tively unstable enzymes in industrial processes involving enzymic reactions. Alternatively, they may be used to isolate the enzyme from a solution thereof thereby assisting purification of the enzyme.

The following abbreviations are used in the examples:

poly-4-acid = N-acryloyl-4-aminosalicylic acid homopolymer

poly-5-acid = N-acryloyl-5-aminosalicylic acid homopolymer

O.D. = Optical density at specified wavelength

Example 1: Interaction of Borate Complexed and Uncomplexed Poly-4-Acid with Lactate Dehydrogenase — Lactate dehydrogenase (10 μl) was added to duplicate samples of free and borate complexed poly-4-acid. The mixtures were immediately centrifuged to separate respectively free and borate complexed poly-4-acid/lactate dehydrogenase complex.

Preparation of Borate Complexed and Uncomplexed Poly-4-Acid — The poly-mers used in the experiment reported above were prepared as follows: sodium 4-aminosalicylate (400 g) and sodium bicarbonate (60 g) were dissolved in distilled water (250 ml) and stirred for one hour. Two additions of acryloyl chloride were made (20 ml and 10 ml), the solution being stirred for one hour after each addition. The solution thus obtained was made slightly acid (pH 4 to 5) by addition of 10 N hydrochloric acid, filtered and washed with distilled

water (500 ml). N-acryloyl-4-aminosalicylic acid was recrystallized from aqueous
ethanol in a yield of 27.0 g. This acid had a melting point of 227° to 229°C
and analyzed as follows: Calculated for $C_{10}H_9O_4N$ (percent): C, 58.0; H, 4.35;
N, 6.76. Found (percent): C, 57.7; H, 4.35; N, 6.65.

The N-acryloyl-4-aminosalicylic acid (15.0 g) and borax (9.36 g) were dissolved
in distilled water (180 ml) and the pH adjusted to 9.0 with 10 N sodium hydroxide. Azobisisobutyronitrile (150 mg) in ethanol (50 ml) was added and
the solution heated at 80°C for 48 hours in a flask fitted with a reflux condenser. The resulting viscous solution was diluted with distilled water (200 ml)
and a white polymer precipitated as a heavy white floc by adding 5 N hydrochloric acid to pH 2. The polymer was washed ten times with 1 l amounts
of distilled water by decantation. The polymer was then rotary evaporated
with methanol to remove any remaining boric acid. The polymer was stored
as a suspension in distilled water (200 ml).

To assess the percentage of water of the polymer, a weighed quantity of filtered polymer was dried over phosphorus pentoxide in vacuo at 60°C. On
drying, a hard, brittle, brown, translucent solid resulted. The filtered polymer contained 93% by weight of water.

The borate complexed polymer was obtained by suspending the filtered polymer (93% H_2O, 500 mg) in an aqueous solution (1.0 ml) of borax (100 mg/ml).
The pH of the mixture was adjusted to 7.0 and the polymer removed by centrifuging.

The uncomplexed polymer was obtained by suspending the filtered polymer
(93% H_2O, 500 mg) in distilled water (1.0 ml). The pH was adjusted to 7.0,
the sample centrifuged and the supernatant removed.

The borate complexed polymer may also be obtained by diluting the viscous
solution supra for polymerizing N-acryloyl-4-aminosalicylic acid (5 g) with
distilled water (70 ml) and then dialyzing the solution for 48 hours against
10 changes (5 liters each) of 0.0005 M borate buffer (pH 7.0).

*Removal of Lactate Dehydrogenase from Poly-4-Acid/Lactate Dehydrogenase
Complex* — A sample of poly-4-acid/lactate dehydrogenase complex prepared
as above was centrifuged and an aliquot of supernatant withdrawn. The remaining supernatant was removed, the polymer washed with distilled water in three
amounts (1.0 ml each) and finally washed with 1 M lactic acid solution (1.0 ml).
Aliquots (500 μl) from each wash were used in a lactate dehydrogenase assay
for pyruvic acid using sodium pyruvate (0.4 μmol) in distilled water (1.0 ml).
Control assays also were carried out using the sodium pyruvate solution (Control A) and using a sodium solution in 1 M lactic acid (0.4 μmol in 1.0 ml)
(Control B). The results are set forth in the following table.

The following results indicate that poly-4-acid takes up lactate dehydrogenase
and that the lactate dehydrogenase can be removed from the complex by washing with strong enzyme substrate, i.e., lactic acid.

	Initial O.D. at 340 nm.	Final O.D. at 340 nm.	Differ- ence
Initial supernatant	1.240	0.630	0.610
Water wash 1	1.200	0.910	0.290
Water wash 2	1.140	0.940	0.200
Water wash 3	1.100	0.960	0.140
Lactic acid wash	1.100	0.215	0.885
Control A	1.204	0.162	1.042
Control B	1.170	0.136	1.034

It should be noted that in the present case the enzyme/polymer complex has
no enzymic activity although it is a source of the active enzyme. This lack
of activity in the solid enzyme is probably due to the similarity in structure
between the ortho-hydroxy acid grouping in the polymer and the α-hydroxy
acid grouping in lactic acid.

*Example 2: Interaction of Titanium Complexed and Uncomplexed Poly-4-
Acid and Poly-5-Acid with β-glucosidase* — The following samples were coupled
with β-glucosidase in the manner set forth:

  4A   Dried titanium complexed poly-4-acid
  4B   Desiccated titanium complexed poly-4-acid
  4C   Uncomplexed poly-4-acid
  5A   Dried titanium complexed poly-5-acid
  5B   Desiccated titanium complexed poly-5-acid
  5C   Uncomplexed poly-5-acid

Each sample (20 mg) was washed five times (5 minutes each wash) with distilled water. β-Glucosidase (1 mg) in distilled water (5 ml) was added to each
sample and the mixture stirred at about 4°C for 16 hours. The mixtures were
centrifuged and the respective supernatant removed. An aliquot (25 ml) was
taken from the initial enzyme solution and another from each supernatant after
coupling. These aliquots were assayed to determine enzyme take-up by the
polymer. The samples were washed five times (5 minutes each wash) with
distilled water and the supernatant removed. 0.005 M acetate buffer (pH 5)
(2 ml) was added and an aliquot taken for assay.

The aliquots supra were assayed by measuring the release of o-nitrophenyl anion
from a solution of o-nitrophenyl-β-D-glucopyranoside. The results obtained are
set forth in the following table.

Sample	Initial enzyme sol (25 μl aliquot) (Difference	Supernatant after cou- pling (25μl) in O.D. 420 nm.)	Solid
4A	0.80	0.052	2.0
4B	0.80	0.048	>2.0
4C	0.80	0.034	2.0
5A	0.80	0.064	>2.0
5B	0.80	0.050	1.95
5C	0.80	0.062	1.30
Reagent blank	0.014	0.042	0.013

The above results show that almost all of the β-glucosidase in the initial solutions was taken up by the polymer samples.

Example 3: Interaction of Titanium Complexed Poly-4-Acid and Poly-5-Acid with α-Amylase — Samples (20 mg) 4A, 4B, 5A and 5B as defined in Example 2 were each washed five times (2 minutes each wash) with distilled water (5 ml). The final supernatant was removed, α-amylase from *Bacillus subtilis* (1 mg) in distilled water (5 ml) added to each sample, and the mixture stirred for 16 hours at 4°C. The supernatant was removed and the digest washed five times with distilled water (5 ml) and ten times (two minutes each wash) with 0.1 M acetate buffer pH 5 (5 ml).

The solid phase enzymes were then suspended in 0.005 M acetate buffer pH 5 (2 ml) and stirred until a homogeneous suspension was obtained. Aliquots (1 ml for 4A and 5A and 0.5 ml for 4B and 5B) of the solid phase enzyme were then assayed for starch conversion to maltose. The activity of the solid phase enzymes were as follows:

	Activity Units*/Mg
4A	2.56
4B	2.27
5A	1.51
5B	4.44

*One α-amylase unit was taken to be that which liberated reducing sugar equivalent to 1 μmol of maltose at 20°C in 1 minute

The above results indicate that α-amylase is complexed by titanium complexed poly-4-acid and by titanium complexed poly-5-acid and that the solid phase enzymes thus obtained have significant enzymic activity.

Basic Anionic Exchange Cellulose

Commercially, most glucose is produced by the hydrolysis of cornstarch. Although fructose can be prepared from sucrose by enzymic conversion, it is more economically attractive to produce fructose from glucose, and in the production of certain corn syrup products, a portion of the glucose is normally converted to fructose. The importance of this conversion of glucose to fructose in corn syrup production lies in the fact that fructose not only is substantially sweeter than glucose, but also is very hygroscopic and difficult to crystallize. Therefore, a convenient means for increasing the conversion of glucose to fructose would find much use in commercial practice for the production of corn syrup.

In the procedures of *T. Sipos; U.S. Patent 3,708,397; January 2, 1973; assigned to Baxter Laboratories, Inc.,* an immobilized glucose isomerase enzyme composition having improved stability to heat and capable of producing increased conversion of glucose to fructose is prepared by thoroughly mixing glucose isomerase, such as obtained from the fermented growth product of *Streptomyces phaechromogenes* and *Lactobacillus brevis,* with basic anionic exchange cellulose in aqueous buffer solution at pH 7 to 10 and recovering the enzyme complex from the reaction mixture.

It has been found that the glucose isomerase enzyme complex exhibits excellent heat stability and can be used repeatedly (e.g., about 5 to 10 times) before 50% of its enzyme activity is lost at 70°C. In conventional practice by comparison, the unbound, soluble enzyme loses substantially all of its activity in only one conversion. Moreover, it has been found that in the matrix supported form the pH optimum of the glucose isomerase changes to one unit lower and the percent conversion of glucose to fructose substantially increases. In general, from about 12% to about 15% more of the available glucose in syrups can be converted to fructose per conversion by employment of the enzyme complex instead of the unbound, soluble enzyme. This improved stability to heat and increased yield of glucose conversion represent outstanding advantages.

In the preparation of the immobilized glucose isomerase enzyme it has been found to be important to use only the basic anion exchange celluloses. By the term basic anion exchange cellulose is meant any cellulose containing basic anion exchange groups bonded to the cellulose molecule and capable of entering into an exchange reaction with anionic groups of other compounds. Preferably, the cellulose is an alpha-cellulose such as cotton, wood pulp, paper, or cotton cloth. In particular, these anion exchange cellulose compounds include the di- and triethylaminoethylated celluloses, e.g., DEAE-cellulose and TEAE-cellulose, and the cellulose derivatives of epichlorohydrin and triethanolamine, e.g., ECTEOLA-cellulose.

In general, the preparation of the immobilized glucose isomerase enzyme complex is achieved by mixing the enzyme or enzyme preparation with the basic anion exchange cellulose in aqueous buffer at a pH of from about 7 to about 10, and preferably at about pH 8. After thoroughly mixing these components, at normal temperature (ca. 25°C) or other convenient temperature conditions, the resulting enzyme complex is recovered such as by filtration, centrifugation and the like separation procedures and then preferably washed or sparged with additional buffer. The filtered and washed enzyme complex can be used in its wet form as is, or the cake can be dried by conventional protein or enzyme drying techniques, such as by shelf, rotary drum or spray drying, but preferably, freeze drying.

Example 1: A 1,000 lb culture growing of *Streptomyces phaechromogenes* is provided by inoculating a medium containing 1% D-xylose, 0.3% K_2HPO_4, 0.1% $MgSO_4 \cdot 7H_2O$, 0.02% $CoCl_2 \cdot 6H_2O$, 1% peptone, 3% wheat bran and 2% corn steep liquor, with *Streptomyces phaechromogenes* from an agar slant, and fermenting 48 hours at 30°C. The crude growth product contains 125,000 glucose isomerase units and is worked up to yield a clear supernatant containing 50,000 glucose isomerase units by passing the ferment through a 20 mesh screen to remove residual bran, adding 4% toluene and then 4% filter aid, and filtering. The clear supernatant is buffered with 0.02 M tris buffer, pH 8.5, and then 500 grams of preequilibrated, wet DEAE-cellulose is added.

The pH of the medium is reduced to about pH 8.0 by the addition of the cellulose exchanger. After mixing thoroughly by stirring for one hour, the mixture is filtered, the filter cake is sparged with 0.02 M tris buffer, pH 8.0, and the washed cake is retained as the active glucose isomerase enzyme complex. This complex has excellent heat stability and can be used for the conversion of D-glucose to D-fructose with substantially increased conversion compared to the conversion with the unbound, soluble glucose isomerase.

Example 2: A glucose isomerase preparation from *Lactobacillus brevis* is substituted for the glucose isomerase preparation from *Streptomyces phaechromogenes* of Example 1 and complexed to DEAE-cellulose according to the procedure of that example. A glucose isomerase enzyme complex having excellent heat stability and the ability to provide increased conversion of glucose to fructose substantially similar to that of the glucose isomerase enzyme complex of Example 1 is obtained. In this particular example, the fermentation media contains 0.04% $MnSO_4 \cdot 4H_2O$, 0.01% $MgSO_4 \cdot 7H_2O$, 0.01% $CoCl_2 \cdot 6H_2O$, 1% D-xylose and 0.1% D-glucose, and incubation is conducted at 37°C for 24 hours.

Example 3: Example 1 is repeated except that an equivalent amount of ECTEOLA-cellulose exchanger instead of DEAE-cellulose exchanger is used for preparation of the immobilized enzyme complex. An immobilized glucose isomerase enzyme product of substantially similar excellent heat stability and ability to convert glucose to fructose is obtained.

Additional Anionic Exchange Cellulose Supports

M.A. Mitz and S.S. Yanari; U.S. Patent 3,126,324; March 24, 1964; assigned to Armour-Pharmaceutical have described a catalase composition which, while being in an insoluble form, demonstrates catalase activity by its ability to hydrolyze peroxides.

In these methods, oxygen gas can be generated by a chemical reaction which involves contacting a peroxide substance with a water-insoluble salt having a catalase component and a cellulose component containing bonded anion exchange groups. Water-insoluble salt refers to a complex formed by the reaction of the acidic groups of catalase and the basic groups of the cellulose anion exchanger. This catalase complex maintains its peroxide-hydrolyzing activity while the catalase component thereof is immobilized. The catalase activity was stabilized when provided as the water-insoluble salt of the cellulose anion exchanger. It had been expected that hydrogen peroxide would competitively displace catalase from the cellulose anion exchanger, whereby the enzyme would be eluted in soluble form.

Hydrogen peroxide is apparently hydrolyzed by the catalase so rapidly that displacement of the enzyme from the cellulose anion exchanger does not occur. However, the catalase can be eluted from the cellulose anion exchanger by conventional eluants such as solutions of sodium chloride and sodium phosphate. Consequently, in this oxygen-generating process, the catalase-cellulose complex reaction system may be recharged without disrupting the operation by eluting inactivated catalase from the cellulose anion exchanger and regenerating the reaction system by adsorbing new catalase thereon.

Sterile oxygen gas and sterile water can be produced simultaneously by a chemical method which involves contacting a peroxide substance with a water-insoluble salt having a catalase component and a cellulose component containing chemically bonded anion exchange groups. In this process the enzymatic hydrolysis of peroxide by the catalase complex results in the evolution of sterile oxygen under pressure, while the residual liquid obtained from this oxygen-generating system consists of sterile water.

This water-insoluble catalase salt, including a catalase component and a cellulose component, may be obtained by reacting a catalase substance with a cellulose anion exchanger, according to conventional ion exchange procedures.

Example 1: A cellulose anion exchanger was prepared by the following method: 2-chlorotriethylamine hydrochloride, in the amount of 10 grams was dissolved in 100 ml of distilled water contained in a one liter beaker. The resulting solution was mixed with 20 grams of absorbent cotton. This mixture was heated in an oven at a temperature of 100°C for a period of 18 hours. Thereafter, the dried cellulose amine hydrochloride material was covered with hot, saturated sodium hydroxide solution, and after 15 minutes the excess alkali was separated by dilution and centrifugation. The cellulose precipitate thereby obtained was washed with distilled water until it was the same pH as the distilled water.

The exchange properties of this cellulose anion exchanger were analyzed by mixing 10 ml of 0.5 M sodium chloride with 100 mg of the exchanger. The pH change thereby produced was from pH 7 to pH 9.6.

Example 2: A few drops of a solution containing crystallized catalase was mixed with 1 gram of the cellulose anion exchanger. The cellulose material had been charged into a glass cylinder, and consequently the dispersion of the catalase through the cellulose anion exchanger could be observed visually due to the natural green color of catalase. Washing the adsorbed catalase failed to mobilize the green color. The catalase was displaced from the cellulose anion exchanger with a dilute aqueous solution of sodium chloride. Also, a dilute solution of sodium phosphate was able to competitively elute the catalase from the cellulose anion exchanger.

ENTRAPMENT

Polyacrylamide Gel

Enzyme gels eliminate or reduce many of the disadvantages inherently associated with soluble enzyme systems. In the enzyme gel, the enzyme is entrapped within a polymer matrix having water colloidally dispersed therein, thereby allowing repeated use of a particular enzyme for different analyses, resulting in greater economy and reduction of error. Furthermore, noncompatible enzyme reactions can be effectively coupled by initiating each reaction sequentially in a series of columns containing different enzyme gels. Therefore, by placing several enzyme gel systems on a flowing stream, each with its own product detection system, several substrates can be analyzed simply and conveniently in a single sample. Thus, the use of enzyme gels promises to extend applications of enzymes and facilitate automation of complex analytical procedures.

In the process of *G.P. Hicks and S.J. Updike; U.S. Patent 3,788,950; January 29, 1974; assigned to E. I. du Pont de Nemours and Company,* a substantially rigid enzyme gel is prepared by polymerizing at least a monomer in the presence of an enzyme in an aqueous solution. The gel is then shaped, mechanically or otherwise, into a desired configuration.

The enzyme gel may be incorporated in a method of chemical analysis comprising contacting the enzyme gel with a substrate solution and determining a characteristic of the resultant enzymatic reaction by monitoring the concentration of at least one reactant or product of the reaction.

The initial requirement for the analytical application of an enzyme gel is that the insoluble enzyme must be prepared in a stable, reproducible reagent form, which can be used in a well-defined configuration for activity measurements. One approach to this problem is the preparation of a stable enzyme gel which, preferably is lyophilized and has a uniform particle size.

The process of preparing the enzyme gel involves polymerizing a monomer compound in the presence of an enzyme while in an aqueous solvent. The compound must be polymerizable, either alone or with other compounds, into a substantially rigid gel and must be present in sufficient quantity to form such a gel. One preferred type of gel has a cross-linked structure. This may result from polymerizing the compound by itself, or with others.

The polymerization may be initiated by catalyzing the solution with chemicals or light. It is also possible for the dissolved enzyme to produce a suitable catalyst.

Depending upon the monomers, enzymes, and catalysts, it may be necessary to adjust the pH prior to catalyzing the reaction or even dissolving the enzyme. The pH range of about 7 to 9 is generally satisfactory.

One advantage of this method of forming an enzyme gel is that individual enzyme molecules are homogeneously dispersed throughout the gel. This results in greater enzyme accessibility to the substrate and faster reaction rates. However, it requires that the gel pore size be sufficiently small to retain these individual enzyme molecules. A pore size of 35 A will retain most enzymes.

An additional advantage of this method is that only aqueous solvents are used. Most enzymes denature when placed in a nonaqueous solvent, whereas very few denature in water. This method avoids placing enzymes in a destructive environment.

In making this gel, stock solution of monomer compound is made by dissolving 40 grams of acrylamide in 100 ml of 0.1 M phosphate buffer, pH 7.4 and a cross-linking reagent is prepared by dissolving 2.3 grams of N,N-methylenebisacrylamide in 100 ml of 0.1 M phosphate buffer, pH 7.4 (both solutions are stored at 5°C). Gels are then prepared by mixing, in the desired proportions, the aforementioned solutions together with an enzyme solution.

Since oxygen inhibits the copolymerization that occurs in this system, it is preferable to deoxygenate the reagent mixture by purging with an inert gas, such as nitrogen, or other means before the addition of the enzyme solution. To catalyze the photopolymerization, riboflavin or potassium persulfate are preferably added. Upon photocatalysis, which may be conveniently performed, with a Number 2 photo floodlamp, the reaction should be complete within two to fifteen minutes, the end point being defined as the time taken for the gel to reach maximum opacity. To reduce the effect of heat denaturation of the enzyme during the exothermic reaction, the reaction container may be placed in an ice bath or other appropriate cooling means during the polymerization.

The catalyst best suited for polymerization of acrylics depends upon the gel composition. In general, mixtures with a high percentage of monomer polymerize more effectively with persulfate as a catalyst, while solutions with a higher percentage of cross-linking agent polymerize better with riboflavin and a photocatalyst. Gels can also be polymerized with no persulfate or riboflavin with a highly purified oxidase, and exposure to room light. The oxidase reaction consumes oxygen and generates H_2O_2 which apparently is sufficient for catalysis.

Various other catalysts and buffer solutions may be used (with corresponding changes in reaction conditions) to polymerize the reactants to the final enzyme gel, e.g., catalysts, tetramethylenediamine and ammonium persulfate, with tris buffer, pH 8.6. The primary requisite that must be met is that the polymerization occur under conditions which do not inhibit final enzyme activity.

The pore size of an acrylamide gel is known to depend upon the concentration of the monomer compound prior to polymerization. Measured pore diameters vary from 17 A for a 5% w/v gel to 7 A for a 35% gel. Radioactively labelled albumin (rather than an enzyme) was incorporated in a gel formulated from 4.45% acrylamide and 0.55% N,N-methylenebisacrylamide. After 60 days, the gel retained approximately 40% of the originally dissolved albumin. This represented 75% of the albumin remaining after the initial washing of the gel. These results assume increased significance since enzyme molecules are generally much larger than albumin and would be expected to display increased retention times.

After the polymerization described above, the resulting block of polymerized enzyme gel may be mechanically dispersed into particles by, for example, first passing it through a number 13 syringe needle and then through a number 16 syringe needle to break the gel into even smaller particles. The suspension of gel particles is then washed, subjected to lyophilization, and sieved to a size between 20 and 40 mesh and stored in a desiccator at 5°C.

This 20 to 40 mesh enzyme gel is generically referred to as E-G 20-40. Other configurations which may be used include blocks, strings, rods, coatings, and tubes.

The technique of preparing E-G 20-40 enzyme gel particles may be applied to any compatible enzyme, e.g., glucose oxidase (GO) catalase, lactic dehydrogenase (LDH), amino acid oxidase, glutamic dehydrogenase, etc.

A series of E-G 20-40 enzyme gel particles having the same enzyme concentration (10 mg of lactic dehydrogenase per 100 ml of gel) but with different compositions of monomer and cross-linking agent were prepared to determine empirically the conditions for optimizing the two properties most useful in flowing stream systems, that is, mechanical rigidity and activity. All gels were prepared as previously indicated except that the ratios of the monomer and cross-linking reagents were varied. More dilute gels for a particular ratio were obtained by diluting the mixture of monomer and cross-linking agents with 0.1 M phosphate buffer, pH 7.4, before adding the enzyme and catalyst.

A list of gel compositions and properties are shown below. The relative activities indicated there were based on the preferred gel composition listed in column 1 of the table being 100.

Characteristics and Activity of Enzyme-Gel Particles Prepared with Different Gel Recipes

Total grams monomer plus cross linking agent per 100 ml. of monomer plus crosslinking agent solutions	8.2	[1] 5.0	[1] 5.0	11.1	14.7	[1] 5.0	5.8	[1] 5.0	4.1	3.2
Percent acrylamide monomer	81	90	95	90	95	81	32	32	49	68
Percent N,N-methylenebisacrylamide crosslinking agent	19	10	5	10	5	19	68	68	51	32
Relative activity of enzyme gel particles	100	66	60	32	8	White, opaque, minimally active, unsatisfactory for use due to poor rigidity which caused columns to plug.				
Mechanical rigidity	[2]	[3]	[2]	[2]	[2]					
Appearance	[4]	[5]	[5]	[4]	[6]					

[1] Diluted with 0.1 M phosphate buffer, pH 7.4, before adding enzyme and catalysts.
[2] Excellent.
[3] Fair.
[4] White, opaque.
[5] White, slightly opaque.
[6] Clear.

Best mechanical rigidity was obtained at higher gel concentrations (higher concentrations of monomer and cross-linking agent in the gel) over the concentration range studied. At any one concentration, increasing the percent of cross-linking agent decreased mechanical rigidity, but favored a higher yield of enzyme gel activity per unit of soluble enzyme activity introduced before polymerization.

Enzymes may be trapped by polymerizing only the cross-linking agent. Such a gel, however, is very soft, sediments slowly, and is unsuitable for use in flowing system applications. On the other hand, high gel concentrations tend to reduce the activity of the gel at a given percentage of cross-linking agent. The most suitable gel material requires both a relatively high concentration of monomer to lend mechanical rigidity and a high concentration of cross-linking agent to achieve the highest possible yield of enzyme gel activity. The concentration of cross-linking agent is limited by its solubility in aqueous solutions, which is less than 3 grams per 100 ml.

The preferred gel was prepared by mixing 1 ml of the acrylamide solution and 4 ml of the N,N-methylenebisacrylamide solution with 1 ml of an enzyme solution (containing from 0.1 to 20 mg of enzyme) and adding approximately 0.03 mg of riboflavin and potassium persulfate.

An insoluble enzyme gel matrix, such as that described on the previous page, can be used as an analytical tool by packing the material into a column. Enzyme gel columns are preferably prepared by equilibrating a dry weight of E-G 20-40 for approximately an hour in a buffer solution, then transferring quantitatively to a column. For example, GO and LDH E-G 20-40 are hydrated prior to transfer in 0.2 M acetate buffer, pH 4.15 and 0.1 M phosphate buffer, pH 7.4, respectively. When the enzyme gel column is perfused with a solution containing the substrates for the enzyme entrapped in the gel, the enzyme reaction is catalyzed by the enzyme gel in the column converting all or a portion of the substrates to a product. The products thus formed emerge from the column free of enzyme activity. The effluent stream containing the formed products can be measured by an appropriate detection system.

In order to be useful for an analytical measurement, the amount of product formed during passage through the enzyme gel column must be proportional to the amount of substrate entering the column, thereby forming the basis for the enzymatic measurement of substrate concentrations. For the measurement of substrates, the substrates may be either quantitatively converted to products during passage through the column, or the conversion may be incomplete with the amount of product formed or reactant consumed being proportional to the amount of substrate entering the column.

Under conditions where the conversion of the substrate is incomplete and, effectively, the rate of a chemical reaction which is catalyzed by the enzyme gel material is being measured, the concentrations of activators and inhibitors for the enzyme entrapped in the gel can also be measured. Thus, an enzyme gel column can be used directly in a system for the determination of substrates, activators and inhibitors, e.g., the continuous determination of glucose based on the subsequent reaction:

$$O_2 \; + \; glucose \; \overset{GO}{\rightleftharpoons} \; H_2O_2 \; + \; gluconic\ acid$$

In this reaction, GO catalyzes the oxidation of glucose to gluconic acid and hydrogen peroxide. Since the samples (substrate solutions) are equilibrated with the oxygen tension of air and glucose is the sought for constituent, GO is the only reagent which must be added to the system for analysis. By providing an enzyme gel and using an electrochemical method for measuring oxygen tension a reagentless determination of glucose is achieved.

Starch, Agar and Carrageenan Gels

In the procedures of *F.L. Aldrich, V.R. Usdin, and B.M. Vasta; U.S. Patent 3,223,593; December 14, 1965; assigned to Melpar, Inc.,* horse serum cholinesterase is entrapped in a gel, with or without subsequent lyophilization, to yield a usable and reusable immobilized horse serum cholinesterase. The immobilized cholinesterase is used by percolating a suitable substrate solution through the gel or lyophilized powder by means of suction or pressure. The substrate solution contains acetylthiocholine, sodium dichloroindophenol and pH 7.4 buffer. The entrapped cholinesterase catalyzes the hydrolysis of acetylthiocholine to acetic acid and thiocholine, which in turn reacts with the blue sodium dichloroindophenol to yield a colorless product.

When an inhibitor of cholinesterase, such as physostigmine, is added to the substrate solution no reaction takes place and the blue color of the sodium dichloroindophenol remains unchanged, i.e., the procedure does not yield a colorless product. Physostigmine is a reversible inhibitor. It can be washed out of the gel by water, buffer, or substrate solution, whereupon the immobilized-enzyme preparation regains its activity.

Immobilized horse serum cholinesterase is useful in detection devices, since it lowers the quantitative requirement for horse serum cholinesterase by several orders of magnitude, and simplifies the construction of detection devices employing the enzyme. The method can be used in any process in which it is desired to hydrolize specific esters, as acetylcholine, butyrylcholine, or phenylacetate, and to recover the hydrolysis products uncontaminated by enzyme.

Essentially, immobilization of horse serum cholinesterase is achieved by incorporation in gels, such as starches, agars, and carrageenans. The resulting preparation is stable, and enzymatically active. It is capable, when so incorporated, of being inhibited by standard anticholinesterases. The gels can be lyophilized to yield dry powders, in order to minimize storage problems, and these powders have been found to be active.

Example 1: 3.25 grams of Connaught starch and 25 ml of distilled water are heated gently until a clear, viscous sol is formed. Six mg of Worthington horse serum cholinesterase are sprinkled on a glass template. 6 ml of the sol, cooled to 45°C, are then pipetted on top of the enzyme, and mixed gently. A filter paper disc is placed over the partially solidified mix, the template is inverted on a smooth surface and pressed gently to remove air. The template is then placed in a refrigerator to allow complete solidification of the gel.

Example 2: 6 mg of Worthington horse serum cholinesterase are added to a sol, in powdered or dissolved form, the sol being 3.25 grams of gelling material to 25 ml of distilled water. Starches, agars, carrageenans, may be utilized as gelling materials. The mix is poured into a flask, which is rotated and cooled until a large film of gel results.

Cellulosic and Other Fiber-Forming Webs

E.E. Schmitt, R.A. Polistina, and P.S. Forgione; U.S. Patent 3,809,605; May 7, 1974; assigned to American Cyanamid Company have described a liquid laid web comprising cellulosic or other fiber-forming material having a carrier bound enzyme dispersed throughout the interstices thereof in an amount of 5 to 50% by weight.

The catalytic activity of immobilized enzymes can be maintained and the handling of the immobilized enzyme can be made more facile by physically entrapping or dispersing the immobilized enzyme throughout the interstices of fibrous webs i.e., sheets or mats, of cellulosic or other fiber-forming materials.

Enzymes per se have been covalently attached to fibers, but the levels of resultant enzyme activity have been very low. Similarly, papers and glass mats have been saturated with enzyme solutions and then dried to afford an enzymatically active paper; however, the enzyme unfortunately migrates and is even removed when it is rewetted by an aqueous solution of the substrate.

Accordingly, to this process there is provided a reaction arena within which a given substrate can be acted upon by a fixed and constant quantity of enzyme. Enzymatic reactions can take place within the vehicle whereby substrate solutions may be added to these systems without loss or movement of enzyme. A wide variety of enzyme activities can be achieved merely by adjusting the ratio of the cellulosic or other fibers to immobilized enzyme with exact reproduction being the rule rather than the exception.

Substrate solutions capable of undergoing enzymatic transformations may be passed through the webs i.e., sheets and/or mats, wherein a chemical reaction occurs and the pure converted substrate emerges from the device containing the web. On a smaller scale, chemical reactions can be conducted uniformly within webs, since the enzyme concentration is always constant and is not affected by evaporation, concentration gradients, hydrodynamics etc.

The carriers for the enzymes may be polymeric or nonpolymeric. Polymers which may be used as carriers for the enzymes, include such materials as aminoethylated cellulose, acid chlorides of carboxylic or sulfonic acid ion exchange resins, diazotized poly-p-aminostyrene, and the like.

Examples of carbonyl polymeric carriers which are preferred for use include those produced according to any known procedure from such aldehyde monomers as acrolein; α-alkyl acroleins, crotonaldehyde; and the like, alone or in mixture with up to 95% by weight based on the total weight of the copolymer, of each other and/or other copolymerizable monomers known to react therewith such as unsaturated alcohol esters, vinyl cyclic compounds, unsaturated ethers, unsaturated amides, etc.

The polymeric carriers are prepared depending, of course, upon the specific material being used, by rendering the material susceptible to reaction with the enzyme and should be hydrophilic in character. In the case of most carbonyl polymers, for example, the polymer is first made water-soluble by reaction with a suitable solubilizing agent such as a sulfite, a hydrosulfite, a bisulfite, sulfurous acid etc.

After the sulfite treatment, the carbonyl polymer is then made hydrophilic, such as by cross-linking. By the term hydrophilic, as used herein, is meant that the carrier is swellable in or capable of taking up water but is not substantially soluble therein. The materials can contain hydrophobic members or portions provided that they also have hydrophilic portions which function as such when in contact with water. Any cross-linking agent or water-insolubilizing agent can be used for this purpose with such materials as bisdiazobenzidine, bisdiazohexane, N,N'-1,2-phenylenebismaleimide, phenol-2,4-disulfonyl chloride, m-xylylene diisocyanate, epichlorohydrin, p-nitrophenyl chloroacetate, tris[1-(2-methyl)aziridinyl]phosphine oxide, diamines such as 1,6-hexamethylene diamine, dialdehydes such as glutaraldehyde and the like being exemplary.

Insolubilizing the carriers can also be accomplished in a multiplicity of other ways such as by reaction thereof with a polyunsaturated cross-linking agent such as divinyl benzene etc. or any other polyfuntional compound which will cause the formation of a network of polymeric structures via reaction with the carrier through available cross-linking sites. Grafting of the polymer carriers by irradiation thereof with x-rays or γ-rays etc. can also be accomplished to render them hydrophilic. Additionally, the insolubilizing can be effected by reacting the carrier material with such agents as 4-aminophenyl sulfide hydrochloride salt etc.

When the carrier polymer is per se water-soluble, the sulfite reaction need not be conducted and the enzyme can be bound to the carrier immediately after rendering it, the carrier, insoluble, such as by cross-linking, as discussed above. Water-solubilizing and cross-linking need not be accomplished, of course, if the carrier is per se hydrophilic. The basic requirement is that the polymeric carrier having the enzyme bound thereto must be hydrophilic in order that it may be utilized in the enzymatic conversion of substrates to their conversion products.

The binding of the enzyme to the carrier is preferably conducted in the presence of buffers (pH 6.0 to 8.0) and with agitation. The binding is accomplished in the presence of water since most solvents tend to deactivate the enzyme.

The immobilized (bound) enzyme is dispersed throughout the interstices of the web of cellulosic or other fiber-forming material in amounts ranging from about 5% to about 50% by weight based on the weight of the web, the resultant web then having from about 0.1 to about 100.0 units, preferably 0.3 to 90.0 units, of activity thereon per ½ inch section of web of a thickness of from about 3 to about 30 mils.

The term water-laid is used to describe a method by which fibrous webs are formed, whether the vehicle is water or other suitable liquid. For most purposes it is preferred that the starting cellulosic fibers be unsized and generally free of added resins. However, for some purposes, it may be desirable to employ as the web, a porous, high wet strength paper such as may be obtained by incorporating into the web from about 0.5 to 5.0% by weight based on the weight of the fibers, of a thermosetting aminoplast resin such as a urea-formaldehyde resin, a melamine-formaldehyde resin and the like. Such wet strength cellulosic webs are obtained in the conventional way by the use of such a resin applied to the pulp suspensions followed by sheeting and baking at temperatures of 210° to 400°F for periods of about 5 minutes to 1 hour.

The carrier bound enzyme may be added to the web anytime during the production thereof as long as the temperatures employed do not deactivate the enzyme. Therefore, to a beater pulp of paper making fibers of any convenient consistency can be added the bound enzyme. The suspension is then agitated gently to distribute the material uniformly therethrough and the aqueous suspension is then sheeted, preferably at a pH of between 4.5 and 6.0, to form a wet, water-laid web containing the bound enzyme. The web is then dried, preferably in air or under vacuum. Vacuum drying with a desiccant of calcium chloride for 6 days has also proven effective. Drying at elevated temperatures is to be avoided since high temperatures tend to deactivate the enzyme and render the dispersed material useless. In general, it is preferred that less than, about 0.1% of residual water be retained in the final sheet.

By incorporating the bound enzyme within the web as it is being formed, there is obtained by direct engagement of fibers and bound enzyme, an integral mechanical union between the bound enzyme and the web. Excellent porosity and permeability of the web permit circulation of the substrate to be treated by the bound enzyme through the structure. Within the web, the fiber components enmesh to form an integral sheet structure which provides a holding and reinforcing matrix for the bound enzyme dispersed through the web.

By employing the water-laying method, a cohesive web is obtained in most cases without need for any further binding material of the nonfibrous type. If a binder is found necessary, however, colloidal polytetrafluoroethylene aqueous emulsion may be added to the fibers and bound enzyme before the web is cast. Agitation of the slurry causes the colloidal suspension to be broken and the PTFE to be coagulated in the slurry. When the resultant slurry is cast, the PTFE extends throughout the web binding the bound enzyme with the fibers. Other polymeric materials can also be used for this purpose, as discussed briefly above.

Once the webs have reduced to dry sheets or mats etc. and recovered, they may be utilized as such or they may be formed into discs or other shaped articles or otherwise treated to produce the desired object.

Example A: Solubilization of Polyacrolein — To a suitable reaction vessel fitted with stirrer, condenser, thermometer, nitrogen gas inlet and constant temperature bath are added 344 parts of sodium metabisulfite and 2,400 ml of distilled water. The pH of this solution is adjusted to 5.6 with 10 M sodium hydroxide solution and 300 parts of finely divided polyacrolein are added.

The reaction is allowed to continue under a nitrogen blanket at 65°C until a clear, viscous, water-soluble polyacrolein adduct forms. The reaction medium is cooled and stored.

Example B: Cross-Linking of Soluble Polyacrolein — To a suitable glass-lined reaction vessel equipped with stirrer and nitrogen gas inlet are added 2,500 ml of the polyacrolein/bisulfite adduct produced in Example A, above, in 4,000 ml of distilled water. The solution is gently stirred and 300 parts of 1,6-hexamethylenediamine in 400 ml of distilled water are added drop-wise over a 4 hour period. A yellow, cross-linked polymer becomes suspended and is heated to 60°C under a heavy nitrogen blanket for 10 minutes and then cooled to room temperature.

The polymer is filtered through cheesecloth, placed in a Büchner funnel and washed thoroughly with water. The cross-linked polymer is then slurried gently with 10 times its volume of water for 15 to 20 minutes, allowed to stand 20 minutes and is filtered.

The washing procedure is repeated until the pH of the washings are between

6.5 and 7.0. The solid adduct is then slurried gently for 20 minutes with 1 M disodium phosphate adjusted to pH 6.5 and washed with distilled water.

Example C: Binding of Enzyme to Cross-Linked Polyacrolein — To a suitable reaction vessel is added 0.625 part of invertase and 50 ml of phosphate buffer (0.02 M; pH 7.4). The solution is allowed to stand in the refrigerator without agitation for 30 minutes. The enzyme is then completely dissolved by stirring. In a separate vessel are slurried 25 parts of the cross-linked polyacrolein of Example B, above (14 mg/g binding capacity) with 50 ml of the same phosphate buffer. After stirring 10 minutes, the pH is readjusted to 7.4 with 0.1 N sodium hydroxide.

The contents of both vessels are then admixed and allowed to stir gently overnight at 15°C. The enzyme adduct is then filtered and washed with copious amounts of deionized water. Consistent binding yields of 75 to 97% are achieved.

Example 1: A 50/50 Albacel/Astracel pulp (concentration 2.6 g/100 cc) is washed with water and subsequently with methanol to remove any residual sulfite and dried. To 1.3 parts of pulp are added 5.0 parts of the wet bound enzyme adduct (100 units/g) of Example C, above, in a suitable blending vessel.

The ingredients are blended for 5 minutes, ice being added to the mixture to prevent heat build-up. The blended slurry is then processed into a paper mat about the thickness of standard filter paper and of six inch diameter on a British Hand Sheet Mold. The paper is then dried in vacuo over a desiccant for 16 hours. The resultant dry sheet yields 236 ½ inch discs with approximately 0.6 unit of activity each.

Treatment of the enzyme disc with 50 μl of 100 mg percent sucrose solution yields an equimolecular mixture of glucose and fructose after 5 minutes of contact time. This is determined by extraction of the converted substrate with water followed by chemical analysis using 3,5-dinitrosalicylic acid.

Example 2: The procedure of Example 1 is again followed except that the enzyme bound as described in Example C is glucose oxidase. The resultant sheet is wetted with a oxygenated, aqueous glucose solution thereby yielding hydrogen peroxide in good quantity.

Example 3: The procedure of Example 1 is again followed except that the enzyme bound as described in Example C is glucoamylase. The recovered sheet is then contacted with a 10% solution of soluble starch to produce an excellent yield of glucose.

Example 4: 10 parts of poly(methylvinyl ketone) are treated with 80 parts of water containing 8 parts of sodium bisulfite at pH 5.8 at 65°C for 18 hours over a nitrogen atmosphere. The resulting bisulfite adduct is then treated with 1.6 parts of ethylene diamine dissolved in 16 parts of water, with stirring for 25 minutes at 65°C. The hydrophilic polymer is dried and dispersed into cellulose fibers as set forth in Example 1.

After contacting with invertase (k=0.6) at 18°C for 16 hours, the cellulose sheet containing the bound enzyme polymer adduct is used to treat sucrose solution as in Example 1. The results are determined to be excellent with respect to the quantity of yield per quantity of charge.

Examples 5 through 12: Following the procedures of either Example 1 or Example 4, except that various copolymers of other homopolymers are substituted for the polymers utilized therein, bound invertase compositions are prepared, dispersed throughout the interstices of a cellulose sheet, formed into appropriate shapes and used to convert sucrose to invert sugar. In each instance, the sheets catalyzed the conversion to glucose and fructose. The results are set forth in the table below.

Example	Carrier
5	Acrolein-β-allyloxyethanol, 80:20
6	Acrolein-methylvinyl ketone, 45:55 (MW 1,400)
7	Poly(vinylethyl ketone)
8	Poly(isopropenylmethyl ketone)
9	Acrolein-ethyl acrylate-styrene, 25:20:55
10	Methylvinyl ketone-vinyl acetate, 5:95
11	α-Methyl acrolein-methyl methacrylate, 50:50
12	Acrolein-vinyl acetate, 60:40

Example 13: 10 parts of wet (~10% solids) cross-linked polyacrolein-sodium bisulfite addition complex dispersed in kraft pulp (prepared as in Example 1) are suspended in 40 parts of water at pH 3.5 and reacted with 0.110 part of crystalline trypsin, dissolved in 4 parts of water. The reaction mixture is stirred for 18 hours at 10°C and at the end of this time the resulting covalently bound trypsin adduct slurry is washed until free of unbound enzyme.

Analysis of the washings at 280 mμ indicates that 72.8% of the enzyme is bound to the polymer. The slurry is then formed into a sheet which when packed in a column utilizing benzoyl arginine ethyl ester (BAEE) substrate, is shown to be effective catalytically after 18 days of continual use.

Example 14: 2.5 parts of poly(isopropenylmethyl ketone) are stirred with 25 parts of water containing 2.1 parts of potassium metabisulfite at pH 5.8 for 16 hours at 75°C over a nitrogen atmosphere. At the end of this time, 1.3 parts of 2-(2-aminoethyl)-5(6)-aminoethylbicyclo[2.2.1]heptane in 5 parts of water are added to the above reaction mixture and stirred at 80°C for 25 minutes. The hydrophilic adduct which results is washed to neutral pH and filtered.

The wet cake is dispersed throughout the fibers of kraft pulp as in Example 13 and then suspended in 30 parts of water and reacted with 0.025 part of glucose oxidase dissolved in 2 parts of water at 10°C for 18 hours. The resultant covalently bound enzyme slurry is washed free of unbound glucose oxidase with cold distilled water, filtered, and formed into a sheet. The results are similar to those shown in Example 1.

Bound Enzyme in a Reticulated Cellular Material

P.S. Forgione, R.A. Polistina, and E. Emil; U.S. Patent 3,791,927; February 12,

1974; assigned to American Cyanamid Company have found that the catalytic activity of enzymes bound to carriers can be maintained and the channelling and compacting of the bound enzyme can be prevented or substantially reduced by entrapping the bound enzyme in a reticulated cellular material.

Example 1: 13.7 parts of the filtered wet cake of bound invertase (13,650 units of activity) are manually worked into a 3" x 6" x ½" sheet of a commercially available reticulated polyurethane foam having 30 pores per linear inch. The enzyme laden sheet then is backed with an one-eighth inch thick sheet of similar dimensions but having 900 pores per linear inch in order to prevent leaking of the enzyme adduct, rolled and forced into a glass contact tube.

The packed column thus prepared is then continuously charged with a 60% sucrose solution. After 740 hours the conversion of sucrose to invert sugar drops from 75% to only 72% at a flow rate of 1.1 ml/min.

Example 2: 10 parts of poly(methylvinyl ketone) are treated with 80 parts of water containing 8 parts of sodium bisulfite at pH 5.8 at 65°C for 18 hours over a nitrogen atmosphere. The resulting bisulfite adduct is then treated with 1.6 parts of ethylene diamine dissolved in 16 parts of water, with stirring, for 25 minutes at 65°C. The hydrophilic polymer is then washed to neutral pH and filtered. The wet cake is suspended in 55 parts of water and reacted with 4 ml of technical grade invertase (k=0.6) at 18°C for 16 hours.

The resulting covalently bound enzyme polymer adduct is washed free of unbound invertase with distilled water and filtered (18.2 g wet). The adduct is separated in two portions. One portion is entrapped as in Example 1 in a foam and both portions are packed in columns and contacted with 10% aqueous sucrose solution for 280 hours. The entrapped adduct affords 80% conversion to invert sugar while the nonentrapped adduct yields less than 5% conversion at the same flow rate.

Following the procedures of Example 1 or Example 2 except that various copolymers or other homopolymers are substituted for the polymers utilized therein, bound invertase compositions are prepared, entrapped, contacted with a 15% aqueous sucrose solution and used in a packed column for the conversion of sucrose to invert sugar. The results are set forth in the table below.

Carbonyl Polymer	No. Hours Without Need for Repacking or Substantial Agitation of Packed Column
Acrolein-allyl glycolate (80/20)	400
Acrolein-methylvinyl ketone (45/55) (m.w. 1400)	580
Poly(isopropenylmethyl ketone)	300
Acrolein-ethyl acrylate-styrene (25/20/55)	725
Acrolein-glycidyl methacrylate (40/60)	420
Acrolein-butadiene (50/50)	470
Acrolein-vinyl acetate (60/40)	600

Bound Enzyme Within Polytetrafluoroethylene

P.S. Forgione and R.A. Polistina; U.S. Patent 3,766,013; October 16, 1973 assigned to American Cyanamid Company have found that the catalytic activity of enzymes bound to carriers can be maintained and the channelling and compacting of the bound enzyme can be prevented or substantially reduced by physically dispersing the bound enzyme throughout the interstices of the network of interconnected, interwoven fibers of polytetrafluoroethylene. By carrier bound enzyme is meant the product produced by binding an enzyme to a carrier by any of four mechanisms: adsorption, inclusion inside the lattice of the carrier, covalent binding and covalent cross-linking.

Carrier is mixed with the polytetrafluoroethylene and an inert, solid, water-soluble component or material and the resultant mixture is then subjected to high shear in order to fibrillate the PTFE. The purpose of the inert, solid, water-soluble component in the mixture is to enable the enzyme to ultimately be contacted with as many available surfaces of the carrier as possible.

The PTFE may be mixed with the carrier and the water-soluble material in the form of a powder or latex. The fibrillation of the PTFE is produced by subjecting the PTFE to a shearing action whereby the PTFE tends to become fibrous in consistency and is present in the form of finely divided microfibrous and sub-microfibrous particles having diameters ranging from about 100 A up to about 2 microns.

Once the carrier-PTFE-water-soluble material mixture has been treated such that the PTFE becomes fibrillated, the carrier and water-soluble material being primarily dispersed throughout the interstices of the interconnected, interwoven fibers of the PTFE as a result, the inert, water-soluble material is extracted, washed or otherwise removed from the resultant system by contacting the system with water, at a temperature ranging from about 0° to 150°C. In this manner, the carrier surfaces are more extensively exposed by the voids left by the washed-out solid and are therefore more susceptible to contact and ultimate binding with the enzyme desired. Under normal conditions, from about 50 to 100% of the water-soluble material can be extracted.

The binding of the enzyme to the carrier after fibrillation of the PTFE and removal of the water-soluble material is almost always necessary. In this manner, the enzyme is not destroyed by the high temperatures required during the fibrillation process. If an enzyme is to be used however, which can withstand such high temperatures and remain active, of course, binding thereof to the carrier may be accomplished before fibrillation.

When polymeric carriers are being used, especially those which must be first modified before the enzyme may be bound thereto, it is also possible to effect all modifications of the polymer carrier after the fibrillation process is terminated since the availability of reaction sites in the carrier due to the water-soluble material extraction allows modification, and consequently effective enzyme bonding, to be conducted.

If desired, once the enzyme has been bound to the carrier which is dispersed throughout the fibrillated PTFE, the resultant composition is formed into an appropriate configuration for use. For example, the composition may be formed into sheets from which pellets etc. may be punched, the composition may be extruded into rods etc. which are then pelletized as is known in the art, and the like.

The fibrillated PTFE having the carrier bound enzyme dispersed therethrough can be packed in a suitable reaction column while still retaining the catalytic activity of the enzyme. Thus, for example, polymer bound invertase can be dispersed throughout the interstices of polytetrafluoroethylene which has been fibrillated, packed in a suitable column and can then be utilized to continuously convert sucrose to invert sugar.

The catalytic activity of the invertase has thereby been maintained and the need for repacking or continuously agitating the column so as to allow continual and complete contact of the sugar with the enzyme is done away with.

Example 1: 0.88 part of 1,6-hexane diamine, containing 3 parts of water, is slowly added, with stirring, to 44 parts of a 10% solution of polyacrolein-sodium bisulfite adduct (MW 80,000). The mixture is then heated to 85°C, for 20 minutes and the yellow hydrophilic product which forms is then washed until neutral with distilled water and filtered.

The wet hydrophilic adduct thus obtained is divided into two equal portions, one of which is suspended in 50 parts of water and reacted with 0.104 part of invertase (twice recrystallized) which had been first dissolved in 4 parts of water. The enzyme reaction mixture (pH 6.5) is gently stirred for 18 hours at 10°C and the resulting hydrophilic covalently bound enzyme polymer adduct is then washed free of unreacted enzyme.

Assay of the hydrophilic enzyme-polymer adduct shows high activity with sucrose solution, in the form of the filtered wet cake. This cake is packed in a reaction column and a sucrose solution is continuously charged to the column. After 72 hours of continual operation, the bound enzyme is channelled to such an extent that conversion of the sucrose to invert sugar drops from 92 to 60%.

Example 2: The remaining portion of wet adduct of Example 1 is dried under vacuum at 40°C for about 10 hours to remove surface water. The adduct then still comprises about 80% water. The adduct is then mixed with 0.5 weight percent of polytetrafluoroethylene (PTFE) emulsion, i.e., an amount equivalent to about 2.5% by weight of dry olefin polymer and 50% of fine sodium chloride.

The mixture is then milled at 190° to 210°F on a preheated rubber mill in order to drive off excess water and produce complete fibrillation of the PTFE which thereby binds the mass into a coherent calendered sheet. Pellets of one-fourth inch diameter are then punched from the sheet and extracted with water for about 10 hours. The pellets are then impregnated with a solution of invertase (0.104 part in 4 parts of water) by soaking at 10°C for 24 hours at pH 6.8.

The pellets are then packed into a reaction column and a sucrose solution is continually charged thereto as in Example 1. After 850 hours, the conversion of sucrose to invert sugar drops from 86 to 79%.

Example 3: 10 parts of wet (~10% solids) cross-linked polyacrolein-sodium bisulfite addition complex dispersed in PTFE (as prepared in Examples 1 and 2) are suspended in 40 parts of water at pH 3.5 and reacted with 0.110 part of crystalline trypsin, dissolved in 4 parts of water. The reaction mixture is stirred for 18 hours at 10°C and at the end of this time the resulting covalently bound trypsin adduct is washed until free of unbound enzyme.

Analysis of the washings at 280 mμ indicates that 78.2% of the enzyme is bound to the polymer. When packed in a column utilizing benzoyl arginine ethyl ester (BAEE) substrate, the enzyme capsules are shown to be effective after 18 days of continual use.

Example 4: 2.5 parts of poly(isopropenylmethyl ketone) are stirred with 25 parts of water containing 2.1 parts of potassium metabisulfite at pH 5.8 for 16 hours at 75°C over a nitrogen atmosphere. At the end of this time, 1.3 parts of 2-(2-aminoethyl)-5(6)-aminoethylbicyclo[2.2.1]heptane in 5 parts of water are added to the above reaction mixture and stirred at 80°C for 25 minutes.

The hydrophilic adduct which results is washed to neutral pH and filtered. The wet cake is dispersed throughout the fibers of PTFE as in Example 2 and then suspended in 30 parts of water and reacted with 0.025 part of glucose oxidase dissolved in 2 parts of water at 10°C for 18 hours. The resultant covalently bound enzyme is washed free of unbound glucose oxidase with cold distilled water and filtered.

Stabilized Intracellular Glucose Isomerase

W.R. Lamm, L.G. Davis, and R.G. Dworschack; U.S. Patent 3,821,082; June 28, 1974; assigned to Standard Brands Incorporated have developed a method of treating cells of microorganisms containing intracellular glucose isomerase to fix or stabilize the glucose isomerase within or on the cells.

This method comprises treating cells of microorganisms containing intracellular glucose isomerase with a salt solution. The treatment may be effected by a variety of methods. For example, an aqueous suspension of the cells may be formed and a salt added to the suspension or the cells may be relatively dry and an aqueous solution of a salt added.

The conditions under which the cells of the microorganisms containing intracellular glucose isomerase are treated with the salt solution to reduce the extractability of the glucose isomerase can vary significantly. Conditions should not be selected which would substantially inactivate the glucose isomerase. The concentration of the salt in the solution, the pH of the treatment medium, the temperature at which the treatment is effected and the particular salt used are interdependent variables and changing any one of the same can substantially alter the efficacy of the treatment.

It is preferred that the treatment be performed so that less than about 50% of the intracellular glucose isomerase is inactivated and the extractability coefficient (hereinafter defined) of the cells is less than about 35%. Most preferably, the treatment is performed so that less than about 15% of the intracellular glucose isomerase is inactivated and the extractability coefficient of the cells is less than about 15%.

The preferred salts for treatment of cells of microorganisms which contain intracellular glucose isomerase are barium chloride, ferric chloride, cobaltous chloride, sodium chloride, chromic nitrate, cupric chloride and mixtures thereof. The most preferred salt is cobaltous chloride. In the case of cobaltous chloride, concentrations in the range of from about 0.005 to about 0.1 molar provide satisfactory results.

The pH at which the treatment is effected may vary significantly, for instance, in the range of from about 6 to about 9. Substantially neutral pH is preferred since there occurs no substantial inactivation of the glucose isomerase at such pH.

The temperature at which the treatment is effected also may vary significantly. Generally, however, temperatures in the range of from about $20°$ to about $85°C$ provide satisfactory results. Temperatures in the range of from about $20°$ to about $40°C$ are preferred.

The cells of microorganisms containing intracellular glucose isomerase are preferably derived from microorganisms of the Streptomyces genus. Particularly preferred microorganisms are *Streptomyces* sp. 21175 and *Streptomyces* sp. 21176.

To enzymatically convert glucose in a glucose-containing solution to fructose, the cells of microorganisms containing fixed or stabilized glucose isomerase may be contacted with the solution in any convenient manner. For instance, the treated cells may be placed in a column and the glucose-containing solution passed therethrough or the cells may be used in a batchwise fashion.

Extractability Coefficient — Cellular material containing fixed or stabilized glucose isomerase is held in an aqueous suspension containing 0.001 mol Co^{++} per liter at $58°C$ and a pH of 6.3. The cellular material is held under these conditions for 24 hours, a portion of the suspension is sonicated at 20 kilocycles by the use of a Branson S75 Sonifier. The sonicated material is centrifuged and the supernatant analyzed for total glucose isomerase activity.

Another portion of the suspension (not sonicated) is filtered and the concentration of the extracted isomerase determined. The concentration of extracted isomerase in the filtrate divided by the total glucose isomerase activity multiplied by 100 is the extractability coefficient of the treated cellular material.

Example 1: This example illustrates the ability of various salts at several concentrations to fix or stabilize intracellular glucose isomerase. *Streptomyces* sp. ATCC 21175 was cultivated in a suitable medium under aerobic conditions for 50 hours in commercial fermentors. A sample of the fermented broth was taken and divided into 100 ml portions. Salts at various concentrations were added to the portions of the fermented broth, the pH was adjusted to 8.5 with sodium hydroxide solution and the portions were then placed in a constant temperature water bath at 30°C for 3 hours.

The pH of the portions was maintained at 8.5 by intermittent additions of a sodium hydroxide solution. Two percent filter aid (Dicalite CP-175) was added to each portion and the portions filtered. The filter cakes were assayed to determine their extractability coefficients and the percent of fixed glucose isomerase activity recovered in the cellular mass. The results are set forth in the table below.

CONDITIONS OF TREATMENT			
Type Salt	Concentration of Salt (Molarity)	Extractability Coefficient (percent)	Isomerase Recovered (percent)
Control (no salt addition)	0	50	49
$BaCl_2$	0.5	5	106
	0.05	51	47
	0.005	50	54
$CaCl_2$	0.5	0	0
	0.05	20	56
	0.005	33	74
$FeCl_3$	0.05	0	0
	0.005	13	74
$FeCl_2$	0.5	0	0
	0.05	0	0
	0.005	23	35
$MnCl_2$	0.5	0	0
	0.05	39	26
	0.005	16	58
$MgCl_2$	0.5	0	0
	0.05	32	70
	0.005	85	11
$NiCl_2$	0.5	0	0
	0.05	3	22
	0.005	53	25
$ZnCl_2$	0.5	0	0
	0.05	2	45
	0.005	12	62
$CoCl_2$	0.05	11	89
NaCl	0.5	22	84
	0.05	62	38
$Cr(NO_3)_3$	0.5	0	0
	0.05	3	74
$CuCl_2$	0.1	5	45
	0.01	9	106
	0.001	37	43
$PbCl_2$	0.1	0	0
	0.01	65	33
	0.001	73	30
$AgNO_3$	0.1	27	71
	0.01	38	78
	0.001	54	50
$HgCl_2$	0.1	72	29
	0.01	57	50
	0.001	60	42

From the above table it is apparent that all the salts tested, except $HgCl_2$, had a beneficial effect on the extractability coefficient of the cells of the microorganism. The concentration of the salt is important since at certain concentrations there was little or no effect on the extractability coefficient. Also, it is seen that although certain salts had a beneficial effect on the extractability coefficient, they substantially inactivated the glucose isomerase.

Example 2: This example illustrates the effect of various concentrations of $CoCl_2 \cdot 6H_2O$ on fixing or stabilizing intracellular glucose isomerase. A fermented broth was prepared in the manner described in Example 1. A sample of the broth was divided into 5 equal portions and various amounts of cobalt chloride salt were dissolved in four of the portions. The fifth portion was not treated and served as a control. The pH of each portion was adjusted to 8.5 with a NaOH solution. The temperature was adjusted to 30°C and held for 4 hours and the pH maintained at 8.5 by intermittent additions of sodium hydroxide. The portions were filtered on a Buchner funnel and the filter cakes assayed to determine their extractability coefficients and the percent of glucose isomerase activity recovered in the cellular mass. The results are set forth in the table below.

CONDITIONS OF TREATMENT

CoCl₂·6H₂O Concentration (Molarity)	Extractability Coefficient (percent)	Isomerase Recovered (percent)
Control (no salt addition)	94	81
0.005	62	96
0.01	33	97
0.05	3	91
0.10	0	81

From the above table it is seen that as the concentration of $CoCl_2 \cdot 6H_2O$ was increased, the extractability coefficient of the treated cells decreased.

Immobilized Microbial Cells

C.K. Lee and M.E. Long; U.S. Patent 3,821,086; June 28, 1974; assigned to R. J. Reynolds Tobacco Company have described a convenient and effective means of adapting enzymes for use in either continuous- or batch-type processes by the use of flocculated microbial cells having the desired enzyme activity.

The method makes separation of the enzyme from the microorganism unnecessary. This, in turn, obviates the need for any enzyme purification or insolubilization procedures. Although this may be applied to any microorganism, it is particularly suited to those organisms producing intracellular enzymes useful in effecting specific transformations of substrates.

The flocculation of microbial cells has received considerable attention in recent years primarily in connection with biological waste treatment and clarification of fermentation liquors. It has now been found that whole microbial cells which have been formed into aggregates by the use of suitable flocculating agents may be used directly in a wide variety of enzymatic processes.

Subjecting flocculated microbial cells to freezing temperatures or to dehydrating conditions imparts a certain rigidity or structural integrity to the flocculated cells without impairing the enzyme activity thus resulting in a material that is particularly adaptable to use in continuous- or batch-type enzymatic processes.

For example, a bed of flocculated microbial cells is able to withstand repeated passage of substrate solutions through the bed with retention of satisfactory flow rates and enzyme activity for extended periods of time.

A variety of flocculating agents are useful in promoting the desired cell aggregation. Such agents include anionic polyelectrolytes such as carboxyl-substituted polyacrylamides, polystyrene sulfonates and polycarboxylic acids; cationic polyelectrolytes such as polyamines, polyethyleneimine and cationic polyacrylamides; polyacids such as polylysine; and mineral hydrocolliods such as activated silicate and colloidal clay. Combinations of two or more flocculants may be used in effecting the desired aggregation.

Selection of the proper flocculating agent in a given instance is easily determined by adding various agents to small samples of the cell-containing broth and comparing the texture and appearance of the cell aggregates formed. The use of filter aids and polymeric adsorbents in conjunction with the flocculating agents may be advantageous in some instances.

For example, amberlite polymeric adsorbents such as acrylic esters and filter aids such as diatomaceous earth and asbestos are effective when added to the broth at the time of flocculation. Amounts of adsorbent or filter aid used may range up to 100% by weight based on the weight of wet cells in the broth.

Addition of flocculant to the cell-containing broth is most conveniently carried out in the form of a solution or suspension. It may also be necessary to adjust the pH of the flocculant solution and/or the broth prior to mixing. This will depend on the microorganism used, the stability of the enzyme of interest and the pH range in which the flocculant is most effective. Flocculation is preferably conducted at ambient temperatures and the quantity of flocculant required is generally 1 to 50% by weight based on the weight of the wet cells contained in the broth.

Agitation of the flocculant-containing fermentation broth must be vigorous enough to cause relatively complete exposure of the cells to the flocculant. Excessive agitation is to be avoided, however, since this tends to break down the size of the aggregates already formed thus defeating the purpose of the flocculant. Careful control of the agitation gives cell aggregates which agglomerate in the fermentor and facilitates virtually complete recovery of the available cells by standard techniques such as decantation, filtration and centrifugation.

The harvested flocculated cells may be used in a contemplated enzymatic process without further treatment; however, the useful life of such cell material in a continuous column operation is reduced considerably due to packing and settling of the material which affects the flow rates adversely. It is preferable, therefore, to subject the harvested flocculated cells to freezing temperatures or to dry the flocculated cells prior to use in an enzymatic process.

Such treatments serve not only to enhance the substrate solution flow rates but also to improve the handling and storage properties of the flocculated cells. The freezing temperatures required in the treatment of the harvested cell material are not particularly critical with $0°C$ and below being satisfactory. Freezing times will, of course, depend on the temperatures used and the bulk of the cell aggregates being frozen.

The consistency of the harvested flocculated cells is such that the material may be extruded into various shapes suitable for use in an enzymatic process. The extruded cell aggregates may be used in conjunction with inert support materials in preparing packed columns for continuous operation or the extrudate may be frozen or dried prior to use.

Drying of the flocculated cells is particularly useful because it permits milling of the dried material and selection of particle sizes by sieving the milled cells. Control of particle size, in turn, insures a uniform bed when the material is packed into columns for continuous flow of substrate solutions through the packed bed.

Drying of the flocculated cells also permits storage of the dried material at ambient temperatures for extended periods of time. The drying procedure may be carried out in any convenient manner such as by forced draft oven, vacuum, drum dryer, etc. provided that the drying temperatures and times used do not result in appreciable denaturation of the desired enzyme. Drying times will vary but it is preferred that the drying process be continued until the material is sufficiently brittle to permit milling.

Example 1: A species of Arthrobacter such as NRRL B-3728 is placed in a sterilized growth medium containing 2% dextrose, 0.3% meat protein, 0.1% yeast extract, 0.6% $(NH_4)_2HPO_4$, 0.2% KH_2PO_4 and 0.01% $MgSO_4 \cdot 7H_2O$. The pH of the resulting medium is 6.9.

Fermentation is allowed to proceed with agitation at 28°C for 60 hours at the end of which time the pH of the broth is approximately 5.5. To this broth is added a 2% (w/v) solution of a suitable polyelectrolyte such as Primafloc C-7 to give a final polyelectrolyte concentration of 0.25% (w/v). The pH of the polyelectrolyte solution is first adjusted to about 5.0 and the fermentation broth is slowly agitated during addition of the polyelectrolyte.

If desired, a filter aid such as Celite 545 is added to the broth at a concentration of about 0.5% (w/v) just prior to addition of the polyelectrolyte solution. Gentle agitation of the broth containing the polyelectrolyte (and filter aid, if added) is continued until stable cell aggregates are formed (approximately 10 minutes). The agitation is then stopped and the cell aggregates are allowed to agglomerate thus allowing most of the clear broth to be separated from the cell material.

The cell material is then collected by vacuum filtration or other suitable means. The harvested cell aggregates are either frozen at –5°C or dried at 55°C before use. Isomerase activity of the harvested aggregates, as determined by standard assay methods, is comparable to the enzymatic activity of cells isolated from the fermentation broth prior to treatment with the flocculant and filter aid. Moreover, an assay of the separated fermentation broth shows no isomerase activity indicating complete recovery of the isomerase by the flocculating technique.

Example 2: This example illustrates the continuous conversion of glucose to fructose using flocculated cells that are obtained by the procedure of Example 1 and stored at –5°C for several hours.

Frozen flocculated cells are immersed in water or a glucose syrup. The cell mass is allowed to thaw and is gently crushed by mechanical means or stirring to give particles of approximately uniform size. This slurry is then subjected to vacuum conditions for approximately 30 minutes to remove occluded gases from the cell particles. A jacketed column of 1 inch diameter is partially filled with water or glucose syrup and the degassed slurry is then poured into the column.

The packed column is heated to 60°C and a 2 M solution of glucose having a pH of 8.0 and containing 0.004 M $MgCl_2 \cdot 6H_2O$ is passed through the column at flow rates which are regulated to give the degree of glucose conversion desired. Conversions ranging up to 50% are attainable by this process.

Example 3: This example illustrates the preparation of dried flocculated cells and the use of the dried cells in continuous column type isomerization of glucose to fructose.

Flocculation is obtained by the combined use of a cationic and an anionic flocculating agent. Five parts of fresh Arthrobacter culture (pH 5.6) are treated with one part of a 1% (w/v) solution of cationic Primafloc C-7. The broth containing the flocculant is slowly stirred until large cell aggregates are formed.

One part of a 1% (w/v) anionic Primafloc A-10 solution is then added and stirring is continued until the large aggregates break into fine particles. The flocculated cells are allowed to agglomerate as in Example 1. The cleared broth is separated and the flocculated cells are recovered by simple vacuum filtration.

The pH of both flocculating solutions is adjusted with caustic before adding to the broth. The cationic Primafloc C-7 solution is adjusted to pH 5 and the anionic Primafloc A-10 is adjusted to pH 7. The final pH of the flocculated culture is about 5.6.

The filtered cell cake is extruded through a suitable die before drying the extrudate in a forced draft oven for about 24 hours at approximately 55°C. The dried cell material is granulated in a mill and sieved to 20 to 40 mesh. The granulated dried cells are packed into a column and glucose isomerization is carried out in the manner described in Example 2.

Example 4: Streptomyces olivaceus, strain NRRL B-3583, is cultivated in a nutrient medium containing 0.7% xylose, 0.3% dextrose, 0.5% beef extract, 0.25% yeast extract, 1.0% peptone, 0.5% NaCl, 0.05% $MgSO_4 \cdot 7H_2O$ and 0.024% $CoCl_2 \cdot 6H_2O$. The fermentation is allowed to proceed at 28°C for 24 hours.

A 2% (w/v) solution of a cationic polyelectrolyte is added to the fermentation broth in amounts equivalent to 0.05% (w/v) of the dry flocculant. The broth is briefly agitated to promote aggregation of the cells. Agitation is then stopped and sedimentation of the flocculated cells is allowed to proceed.

The cell aggregates are harvested in a manner analogous to that described in Example 1. The flocculated cells are dried at 60°C for 24 hours before subjecting the dried cell mass to gentle mechanical crushing action to reduce to particle sizes suitable for column operation. The crushed cell material is poured into a jacketed column partially filled with a 1.0 M glucose solution. The packed column is heated to 70°C and a 1.0 M glucose solution containing 0.04 M $MgCl_2 \cdot 6H_2O$ adjusted to pH 8 is passed through the column to give conversion of glucose to fructose.

Example 5: Aspergillus oryzae is cultivated in a nutrient medium containing wheat bran mixed with additional carbohydrates, minerals and buffering substances. Fermentation is allowed to proceed for 48 hours at 30°C at the end of which time a solution of a suitable polyelectrolyte is added to the broth.

The resulting aggregation of the flocculated cells is aided by gentle agitation. The flocculated cells are harvested by decantation and filtration. The flocculated cells are then placed in a jacketed column and heated to 50°C. A neutral 0.2 M solution of acetyl-D L-methionine containing 5×10^{-4} M Co^{++} is passed through the column to give an effluent containing L-methionine.

MICROENCAPSULATION

Polymers Containing Reactive Carboxylic Acid and/or Anhydride Groups

Methods for binding enzymes to carriers include adsorption, inclusion inside the lattice of the carrier, covalent binding and covalent cross-linking. *P. Salvatore and R.A. Polistina; U.S. Patent 3,730,841; May 1, 1973; assigned to American Cyanamid Company* have found that the catalytic activity of enzymes bound to carriers can be maintained and the channelling and compacting of the bound enzyme can be prevented or substantially reduced by encapsulating the bound enzyme in a coating through which the substrate and its conversion product is permeable. By carrier bound enzyme as used herein, is meant the product produced by binding an enzyme to a carrier by any of the four mechanisms mentioned above.

The compositions which are encapsulated in permeable coatings comprise a water-insoluble, hydrophilic carrier having a catalytically active enzyme bound thereto. Any carrier having these properties may be utilized and the enzyme may be bound thereto in any manner, i.e., those mentioned above.

Those polymers containing reactive carboxylic acid and/or anhydride groups are among carriers useful herein. Polymers which may be used as carriers for the enzymes are such materials as aminoethylated cellulose, diazotized poly-p-aminostyrene, polyacrylamide, and the like.

In the preparation of carrier bound enzyme products, polyacrolein, a water-insoluble polymer which contains some groups with which most enzymes are reactive, must be first contacted with a bisulfite such as sodium bisulfite in order to render it water-soluble. In such a condition, however, the polymer cannot be reacted with an enzyme because recovery of any product is impossible.

Cross-linking of the bisulfite-polymer product, however, renders it gel-like in consistency and effectively hydrophilic so as to allow reaction with the enzyme. As a result, the bisulfite-polyacrolein product is preferably cross-linked with a diamine such as hexamethylene diamine.

The result of these two reactions is believed to be that the bisulfite breaks some of the heterocyclic rings of the polyacrolein creating more enzyme-reactive aldehyde groups. The diamine reacts with some of these aldehyde groups with the formation of —CH=N— linkages between two polymer molecules, thereby cross-linking the polyacrolein.

Reaction of the enzyme, e.g., invertase, forms an adduct or covalent bond between the enzyme and the other available aldehyde groups, and also may result in reaction through the bisulfite groups. The resultant adduct is then comprised of a series of cross-linked groups, free aldehyde groups, heterocyclic bisulfite reaction groups and bound enzyme groups.

The encapsulated carrier bound enzyme can be packed in a suitable reaction column while still retaining the catalytic activity of the enzyme. Thus, for example, polymer bound invertase can be encapsulated in an acrylamide-methylene bisacrylamide coating, packed in a suitable column and can then be utilized to continuously convert sucrose to invert sugar. The catalytic activity of the invertase has thereby been maintained and the need for repacking or continuously agitating the column so as to allow continual and complete contact of the sugar with the enzyme is done away with.

The bound enzymes may be encapsulated by any procedure. Encapsulating machines may be utilized, for example, or, more preferably, the bound enzyme may be coated by polymerization (or other means of production) under known reaction conditions, of the coating material in the presence of particles of the bound enzyme. In this manner, the bound enzyme is coated with the polymer produced during polymerization and the resultant capsules may be recovered by filtration etc. and dried.

The materials which may be used to encapsulate the bound enzymes should be tack-free and insoluble in the substrate or substrate solution or, at least, water-resistant thereto, in order to prevent the washing away and ultimate over exposure of the bound enzyme adduct.

The encapsulation is preferably conducted in the presence of an emulsifying agent such as the sodium salt of dodecyl benzene sulfonate etc. to assure particle retention. In order to maintain the enzyme activity at its peak over the period of time between encapsulation and use, the incorporation of a small amount, i.e., 1 to 15% by weight based on the weight of the enzyme bound, of the substrate which is normally converted to its conversion product by the enzyme into the bound material is effective.

For example, in the case of invertase, a small amount of sucrose would be encapsulated therewith. Particle sizes ranging from about 5 microns to about 200 microns are satisfactory for encapsulation.

Examples of suitable coating materials include the alginates such as sodium alginate etc., the celluloses such as ethyl cellulose, carboxymethyl cellulose, ethyl β-hydroxyethyl cellulose, nitrocellulose (collodion) etc., vinyl polymers such as vinyl pyrrolidone-vinyl acetate polymers, polyamides, e.g., adipic acid-hexamethylenediamine reaction products (nylons), acrylamide-methylenebisacrylamide polymers and the like.

Example 1: 0.88 part of 1,6-hexane diamine, containing 3 parts of water, is slowly added, with stirring, to 44 parts of a 10% solution of polyacrolein-sodium bisulfite adduct (MW 80,000). The mixture is then heated to 85°C for 20 minutes and the yellow hydrophilic product which forms is then washed until neutral with distilled water and filtered.

The wet hydrophilic adduct thus obtained is suspended in 50 parts of water and reacted with 0.104 part of invertase (twice recrystallized) which had been first dissolved in 4 parts of water. The enzyme reaction mixture (pH 6.5) is gently stirred for 18 hours at 10°C and the resulting hydrophilic covalently bound enzyme polymer adduct is then washed free of unreacted enzyme.

Assay of the hydrophilic enzyme-polymer adduct shows high activity with sucrose solution, in the form of the filtered wet cake. This cake is packed in a reaction column and a sucrose solution is continuously charged to the column. After 72 hours of continual operation, the bound enzyme is channelled to such an extent that conversion of sucrose to invert sugar drops from 92 to 60%.

Example 2: A charge of 35 parts of a mixture of 95% acrylamide and 5% methylenebisacrylamide is added to a suitable vessel containing 4,000 parts of water. The resultant solution is purged with nitrogen gas, stirred and the following materials added in the order listed:

(1) 1.75 parts of dimethylaminopropiontrile activator,

(2) 3.5 parts of inhibitor-free styrene (utilized to impart a greater degree of hardness in the acrylamide polymer as is known in the art),

(3) 7.0 parts of the filtered wet cake of bound invertase produced as in Example 1,

(4) 3.5 parts of sucrose (utilized to maintain enzyme activity),

(5) 50.0 parts of a 2.0% aqueous solution of the sodium salt of dodecyl benzene sulfonate, an anionic surfactant, and

(6) 3.5 parts of ammonium persulfate in 20.0 parts of water, a known initiator.

Upon completion of the persulfate addition, the resultant reaction media is stirred at moderate speed for 30 minutes and at 35°C in order to dispense the adduct as small particles of about 15 to 50 microns. Upon completion of the polymerization (about one hour), the resultant encapsulated particles are filtered and washed with large amounts of water.

The capsules containing the bound enzyme are packed into a reaction column and a sucrose solution is continuously charged to the column as in Example 1. After 920 hours, the conversion of sucrose to invert sugar drops from 91 to 88%.

Example 3: Following the procedures of Examples 1 and 2, polyacrolein bound glucose oxidase is prepared and encapsulated. The sucrose of Example 2 is replaced with 10.0 parts of a 40% glucose solution. After a similar length of time in a packed column, the encapsulated adduct retained 98% of its activity. A similar column of nonencapsulated adduct shows channelling and unsatisfactory conversion in 48 hours.

Synthetic Organic Polymers

S. Sternberg, H.J. Bixler, and A.S. Michaels; U.S. Patent 3,639,306; February 1, 1972; assigned to Amicon Corporation have described anisotropic polymer particles comprising a thin skin which forms either a microporous or diffusive barrier layer and a substantially hollow interior which interior, however, comprises sufficient macroporous polymer support structure to substantially enhance the physical strength, particularly the resistance to compression, of the particles.

These particles have excellent utility for encapsulating various materials for controlled release therefrom, for immobilizing some chemical reactants while allowing free access and egress of other reactants or reaction products from the particle, and for effecting separations based on diffusion of different size molecules therethrough.

By proper selection of process reactants and conditions, the outer barrier layer of the particle may have a preselected degree of permeability (or microporosity) as appropriate for the particular application for which the particle is intended. Another important aspect of the particles; one which makes it practical for such beads to be used in many separation processes; is in fact that the transition between the microporous and macroporous sections of the particle (which may be considered in concentric relation to each other) takes place within extraordinarily small geometric space; usually within about 0.5 micron or less. Thus the particles do not act like easily-plugging depth-filters and have an acceptably useful life.

In effect then, particles form a kind of particulate ultrafiltration membrane with the outer barrier skin being the membrane or separating screen and the inner macroporous layer serving the dual purpose of providing a reservoir for core material and providing good support against compression of the particle.

Moreover, the sharp transition between macroporous and microporous layers together with the extreme thinness of the microporous layers allow very small particles to be prepared and thereby allow for larger effective filter areas to be put into a given volume than would otherwise be expected.

Particles have a highly anisotropic, submicroscopically porous barrier skin and are formed of organic film-forming polymers having good mechanical integrity, most advantageously those crystalline and/or glassy thermoplastic polymers. By crystalline and glassy polymers are those materials which possess from about 5 to 90% by weight crystallinity as measured by x-ray diffraction techniques known to the art and/or a glass transition temperature (T_g) of at least about 20°C.

Particularly advantageous are polymers of inherently low water absorptivity, which may be allowed to dry during storage without destroying the beneficial mechanical and processing characteristics of the microporous barrier layer thereof. These particles do not necessarily depend upon any inherent water absorptivity of the polymer for their utility and many of the most desirable polymers are those having water absorptivities of less than about 10% by weight of moisture at 25°C and 100% relative humidity.

The internally reinforced particles are prepared by:

(1) forming a solution of a polymer in an organic solvent,

(2) forming drops of the solution,

(3) contacting the drops with a diluent characterized by a high degree of miscibility with the organic solvent and a sufficiently low degree of compatibility with the casting dope to effect rapid precipitation of the polymer, and

(4) maintaining the diluent in contact with the particle until substantially all the solvent has been replaced with the diluent.

The submicroscopically porous anisotropic particles consist of macroscopically thick particles of porous polymer. The outer surface of this particle is an exceedingly thin, but relatively dense barrier layer of skin of from about 0.1 to 5.0 microns thickness of microporous polymer in which the average pore diameter is in the millimicron range, for example from 1.0 to 500 millimicrons; i.e., about one-tenth to one-hundredth the thickness of the skin.

The balance of the particle structure includes a support structure comprised of a much more coarsely porous polymer structure through which fluid can pass with no significant increase in hydraulic resistance over that offered by the barrier layer. When such a particle is employed as a molecular filter in contact with fluid under pressure, virtually all resistance to fluid flow out of and into the particle is encountered in the skin. Molecules or other dispersoids of dimensions larger than the pores in the skin can be retained within the particle structure.

Because the barrier layer is of such extraordinary thinness, the over-all hydraulic resistance to fluid flow through the barrier skin is very low; that is, the skin displays surprisingly high permeability to fluids. Furthermore, tendency of the skin of such particles to become plugged or fouled by molecules or other dispersants is surprisingly low.

The anisotropic membrane structures can be prepared in the form of particles by the precipitation and leaching of specially formulated polymer solutions under controlled conditions is believed to be a consequence of unusual coaction of diffusion and polymer precipitation phenomena which occur during a controlled leaching process.

However, by proper selection of solvent and diluent, a variety of polymer-gel structures, of controllable porosity and pore fineness can be prepared for a given polymer. In those instances where a single polymer solvent of the desired degree of polymer compatibility to achieve a desired gel structure cannot be found, it is possible to employ mixtures of miscible solvents of differing compatibility with the polymer (both of which are miscible with the diluent) to achieve this purpose.

In order to obtain approximately spherical particles (as is usually preferable both from the point of view of obtaining particles having good packing quality for use in packed columns, good handling qualities, and barrier skins free from imperfections), it is desirable to introduce the droplet of polymer solution into the wash bath at a sufficient velocity to overcome the surface tension of the diluent. This precaution will avoid the flattening of the particles.

Surfactants can be used to reduce the surface tension of the diluent and lessen the optimum impact velocity. Agitation of the diluent is desirable when the particles tend to float therein because substantially all the solvent must be removed from the particle to assure that it will not be softened by residual solvent on storage.

Internal structure of the spheres can be strongly affected by changes in temperature. When the temperature of the diluent bath is too high, as in a boiling water bath, an open central core or hollow section will usually be achieved within the macroporous support structure. In room temperature water, the central core will be more sponge-like in appearance.

Film-forming polymers include the following: polycarbonates, polyvinyl chlorides, polyamides, polysulfones, acrylic resins and the like. Other polymers such as polyurethanes, polyimides, polybenzimidazoles, polyvinyl acetate, aromatic and aliphatic polyethers, and the like may also be utilized.

The large number of copolymers, which can be formed by reacting various proportions of monomers from which the aforesaid list of polymers were synthesized, are also useful for preparing microporous particles according to the process.

In general, preferred polymers are those which exhibit modest levels of crystallinity at ambient temperatures, e.g., between about 5 to 90% crystallinity as measured by x-ray diffraction analysis and/or those which display relatively high glass transition temperatures, (e.g., at least 20°C, and preferably higher). Polymers meeting these requirements, as a rule, yield particles with good mechanical strength, and good long-term stability at elevated temperatures.

Among the many specific polymer-solvent systems which have been found to be useful in forming particles are the following:

System No.	Polymer	Solvent
1	Acrylonitrile (40)-vinyl-chloride (60) copolymer (Dynel).	N,N'-dimethylformamide (DMF).
2	Acrylonitrile (40)-vinyl-chloride (60) copolymer.	Dimethylsulfoxide (DMSO).
3	Acrylonitrile (40)-vinyl-chloride (50) copolymer.	N-methyl-pyrrolidone.
4	Acrylonitrile (40)-vinyl-chloride (50) copolymer.	Dimethylacetamide (DMAC).
5	Polyacrylonitrile	DMF.
6	do	DMAC.
7	Polysulfone	N-methyl pyrrolidone.
8	do	N,N-dimethylpropion-amide.
9	Polyvinylchloride	DMF.
10	do	DMAC.
11	Polyvinylidene chloride	DMF.
12	Polycarbonate	DMF.
13	Polystyrene	DMF.
14	Poly-butyl methacrylate	DMF.
15	Polymethylmethacrylate	DMF.
16	Polysulfone	Cyclohexanone.

(continued)

System No.	Polymer	Solvent
17	Polymer 360	DMAC.
18	do	DMF.
19	do	DMSO.
20	Polyacrylonitrile	70% ZnCl₂ (aqueous).
21	Polycarbonate	DMSO.
22	do	DMAC.
23	do	Tetrahydrothiophene.
24	do	n-Butyrolactone.
25	do	N'N'-diethylformamide.
26	Polyvinylchloride	N'N'-diethylpropionamide.
27	do	n-Butyrolactone.
28	Polymer 360	Tetrahydrothiophene.
29	Dynel	Ethylene carbonate.
30	do	N'N'-diethylpropionamide.
31	do	Tetrahydrothiophene.
32	do	N'N'-diethylformamide.

Other systems which can be utilized include such systems as nylon in solution with phenol, cresol, or formic acid, polyphenylene oxide in DMF; polycarbonate in ethylene carbonate; polyvinyl fluoride in butyrolactone, polyurethane in DMAC.

Usually the pore-structure of the particles can be further modified by the addition of a solution-modifier and/or by further moderate increases in the temperature of the casting and wash operations, and/or by changes in polymer concentration in the casting dope.

Solution modifiers are often advantageously selected to increase the solvating effect on the polymer of the overall solvent system. The use of such a solution modifier will tend to loosen the barrier skin, i.e., increase the size at which a molecule may enter and exit therethrough. By increased solvating effect is meant an increase in compatibility or the degree of proximity to formation of an ideal solution.

Conversely, a solution-modifier which reduces the solvating effect of the overall solvent system tends to decrease the size at which a molecule may enter and exit therethrough. The following will serve to illustrate this with respect to making Dynel particles with water as the diluent and DMF as the primary solvent.

DMF has a solubility parameter $[1/(cal/cc)^{1/2}]$ of 12.1 is a strong to medium hydrogen bonding solvent, and has a dipole moment of 2. Water has a solubility parameter of 23.4, is a strong hydrogen bonding solvent, and has a dipole moment of about 1.8.

Thus a solution modifier used in the process and having a solubility parameter of 10.0, medium hydrogen bonding tendency, and a dipole moment of 2.9 would be expected to decrease the solvating effect on Dynel and thus would tend to tighten the Dynel particle skin. Such is the case with acetone used as a solution modifier, for example in the quality of 5% based on weight of total solvent. Tetrahydrofuran is an example of another such modifier.

On the other hand, a solution modifier having about the same dipole moment as DMF and a strong affinity to water would function more like the ZnCl₂ type of inorganic salt to be discussed below, and has sufficiently greater compatibility with water than DMF to loosen the barrier skin, i.e., increase the mass transfer across the barrier skin under given conditions.

Such is the case with formamide used as a solution-modifier, for example in the quantity of 5% based on the weight of the total solvent. This is in spite of the fact that bare reference to the solubility parameter of formamide would lead one to believe that its use would result in a poorer solvent for Dynel and, consequently, a particle skin having smaller micropores.

In general, a large number of such solution modifiers can be selected for a given polymer-solvent system. The selection can be made, not only from the classical lists of organic solvents, but also from solid organic compounds which may be solubilized in the primary solvents.

Another class of solution modifiers are the inorganic electrolytes dissociable in organic solutions, for example many halides, nitrates and the like. Some such compounds are $FeCl_3$, LiBr, LiCl, $Al(NO_3)_3$, $Cu(NO_3)_2$ and NaCNS and the like.

These materials in solution, tend to have a solvating effect on polar polymers and tend to increase the flux rate attainable with particles formed of solutions in which they are incorporated as solution modifiers. Some such inorganic electrolyte solution modifiers particularly useful in the systems described in the first table include those exemplified by the list in the table below.

System	Solution modifier	System	Solution modifier
1	$ZnCl_2$	2	$ZnCl_2$
1	$FeCl_3$	4	$LiNO_3$
1	LiBr	6	LiCl
1	$Al(NO_3)_3$	11	$ZnCl_2$
1	NaCNS	12	$ZnCl_2$
1	$Cu(NO_3)_2$		

The effect of these salts which act as solvating aids for polymers is usually the opposite when they are incorporated in the diluent.

Example 1: A quantity of 0.59 gram of catalase, an enzyme, was dissolved in 5.9 grams of water. To this was added 54 cc of DMF to form a suitable solvent. Six grams of Dynel was dissolved in this solvent and the resulting Dynel solution was added dropwise to 2,000 cc of a water diluent bath which contained about 0.5 cc of a surfactant. The resulting particles, about 2 to 3 mm in average diameter, were swirled about in the bath for 4 hours during which time they became hard white spheres.

These spheres were put into a 3% aqueous solution of hydrogen peroxide. Evolution of gas bubbles was observed over a period of many hours. Thus it was clear that the enzyme was acting as a catalytic agent, trapped in the particle, to achieve the decomposition of hydrogen peroxide entering the particles. This decomposition is accomplished by the reaction

$$2H_2O_2 \xrightarrow{\text{catalase}} O_2 + 2H_2O$$

The oxygen was able to escape through the barrier skin of the particles. Because the enzyme is locked into the particles, the enzyme itself cannot escape.

Therefore its effectiveness is prolonged. Such catalase-encapsulated material would have utility in a physiological system where unwanted hydrogen peroxide was forming and could be utilized without fear of the enzyme reaching parts of the system in which it could cause damage.

Example 2: Three grams of a polyacrylonitrile fiber (Orlon) was dissolved in thirty grams of an aqueous solution comprising 70% by weight of zinc chloride. After the solution was cooled to about 25°C, 0.2 gram of the enzyme catalase was added to the solution. Thereupon, the solution was added dropwise through a Number 21 hypodermic needle into an acetone bath whereupon particles were formed. After being stirred for about thirty minutes in the acetone, the particles were removed and placed in a water bath for four hours at room temperature to leach out residual acetone and salt.

The residual particles, when immersed in a dilute H_2O_2 solution, within several hours became buoyed by oxygen being formed within and on the surface of the enzyme-encapsulating particles.

Gelatin and Calcium Alginate

J. Shovers and W.E. Sandine; U.S. Patent 3,733,205; May 15, 1973; assigned to Pfizer Inc. have described the coating of diacetyl reductase and reduced nicotinamide adenine dinucleotide; or these enzymes in combination with viable brewers' or bakers' yeast, containing within the cell walls of the yeast the enzyme-co-factor system of diacetyl reductase and reduced nicotinamide adenine dinucleotide, with gelatin or calcium alginate.

The diacetyl content of beer is lowered by the addition of a small but effective amount of one of these coated materials to beer during the latter stages of fermentation or during any of the finishing processes subsequent to fermentation and prior to packaging.

Isolated diacetyl reductase and reduced nicotinamide adenine dinucleotide enzymes are encapsulated with gelatin or calcium alginate. The semipermeable protective coating allows the enzyme-co-enzyme system to operate at the natural acidic pH of beer (3.9 to 4.4) without undergoing loss of activity due to denaturation of diacetyl reductase or irreversible acid hydrolysis of reduced nicotinamide adenine dinucleotide.

Bakers' or brewers' yeast (containing diacetyl reductase and reduced nicotinamide adenine dinucleotide) is incorporated with diacetyl reductase and reduced nicotinamide adenine dinucleotide or with diacetyl reductase alone. The microencapsulating film prevents the bakers' yeast (*Saccharomyces cerevisiae*) or the brewers' yeast (*Saccharomyces carlsbergensis*) from escaping into the beer where it could be considered a wild type yeast by the brewer, while maintaining cell viability.

Furthermore, the protected yeast cells can operate at the naturally acidic pH of beer (3.9 to 4.4) without undergoing denaturation or inhibition of the enzyme systems contained within the cell structure of the yeast.

A high-bloom (high MW) gelatin is employed for the microencapsulation process because of its insolubility in beer. Ratios of materials to be encapsulated to gelatin of from 1:1 to about 9:1 can be successfully utilized but the amount of gelatin is critical only in that it must completely microencapsulate the active components.

For encapsulated materials containing whole yeast cells, complete microencapsulation is readily determined by plating out the material on wort agar. No growth of yeast cells will result from satisfactorily encapsulated material.

Gelatin is added to water and heated to about 40°C until the gelatin has dissolved. The solution is cooled to about 30°C and the pH adjusted to 6.0 with 1 M NaOH. A mixture, in stoichiometric ratio, of diacetyl reductase and reduced nicotinamide adenine dinucleotide is stirred in and the mixture spread uniformly as a thin film on sheets of polyethylene.

After about 24 hours at room temperature, the translucent gelatin-enzyme-cofactor film is peeled from the polyethylene and cut into small pieces approximately 1.0 cm square. The film pieces are stored in a tightly stoppered container at 2° to 5°C.

In an alternate process, the gelatin-enzyme mixture is chilled to about 2° to 10°C until the gelatin has set and the coated mass is macerated with a high shear mixer. The material is tray dried at about 2° to 10°C and then about 22° to 25°C to a final moisture content of about 7 to about 11%, preferably about 8%.

The regeneration of co-factor (reduced nicotinamide adenine dinucleotide) by viable cells of bakers' or brewers' yeast makes possible a mixture of whole yeast cells, diacetyl reductase and the reduction of co-factor to about 0.1 the normal amount or even its elimination entirely.

When mixtures of diacetyl reductase and reduced nicotinamide adenine dinucleotide containing whole yeast cells are encapsulated and recovered as ground or milled material, there is the possibility of sheared surfaces exposing nonencapsulated yeast cells. This is obviated by a coacervation process in which the yeast containing material is dispersed in light mineral oil which is then added in a thin stream with vigorous stirring into an aqueous solution of high-bloom gelatin.

The enzymes and yeast cells are dispersed in the oil medium and each particle is coated with a gelatin-water film. The mixture is chilled to about 10° to 20°C which hardens the gelatin around the yeast cells and enzyme particles with the formation of discrete beadlets. Several liters of cold ethanol are added to partially dehydrate the beadlets. The beadlets are removed by filtration, washed with cold hexane and tray dried or fluid-bed dried at about 5° to 10°C to a final moisture content of about 7 to about 11%, preferably about 8%.

A process for encapsulation in beadlet form involves the use of a water and alcohol insoluble metal alginate. The materials to be encapsulated are dispersed in an aqueous solution of a water-soluble alkali metal alginate such as potassium alginate, or preferably, sodium alginate, the concentration of which is limited to approximately 2% because of viscosity. This is added to a solution of an inorganic salt which is capable of reacting with the soluble metal alginate to form an insoluble metal alginate, and while precipitating out of solution, simultaneously coats the individual yeast cells and enzyme particles with a film of insoluble metal alginate.

For reasons of toxicity, the preferred inorganic salt is calcium chloride which reacts with sodium alginate to form a film of nontoxic calcium alginate around the individual yeast cells. Other inorganic salts which are capable of reacting with sodium alginate to form water and alcohol insoluble alginates which are nontoxic may also be used. Beadlets prepared by this process are tray dried or fluid-bed dried to a final moisture content of about 10 to about 20% preferably about 15%.

Example 1: Ten grams of gelatin are added to 50 ml of distilled water, and the mixture heated to about 40°C until the gelatin is dissolved. The solution is cooled to 30°C and the pH adjusted to about 6.0 with 1 M NaOH. Four hundred milligrams of freeze-dried diacetyl reductase enzyme preparation and 160 mg of reduced nicotinamide adenine dinucleotide are then mixed in, and the mixture spread uniformly as a thin film on sheets of polyethylene.

After about 24 hours at room temperature, the translucent film is peeled from the polyethylene sheets, cut into pieces approximately 1.0 cm square, and stored in a tightly stoppered container at 2° to 5°C.

To 120 ml of beer containing 0.5 ppm w/v of diacetyl is added 100 mg of the gelatin coated enzyme system. After approximately 72 hours at 5° to 7°C, the diacetyl concentration is 0.05 ppm w/v.

Example 2: Four grams of freeze-dried diacetyl reductase enzyme preparation and 1.6 grams of reduced nicotinamide adenine dinucleotide are mixed with 100 grams of a 10% w/v aqueous solution of gelatin at 50° to 70°C, dispersed in a homogenizer for several minutes and chilled in a stainless steel tray in a refrigerator at about 5°C until the gelatin has set.

The coated enzyme mass is broken up, transferred while still cold to a suitable vessel and macerated with a high shear mixer in about a liter of hexane chilled to about 23° to 25°C, and dried to a final moisture content of about 7 to about 11%.

Example 3: Four grams of freeze-dried diacetyl reductase enzyme preparation and 1.6 grams of reduced nicotinamide adenine dinucleotide are dispersed in an aqueous 25% slurry into 200 ml of light mineral oil (USP) at about 30° to 40°C with a propeller-type mixer. With continuous vigorous stirring at 30° to 40°C, 100 grams of a 20% w/w aqueous solution of high-bloom gelatin is added in a thin stream.

The mixture is then chilled to about 10° to 20°C. Approximately 200 ml of cold ethanol is added. After decantation, the beadlets are filtered and washed with cold hexane. The beadlets are tray dried or fluid-bed dried at 5° to 10°C to solids containing about 7 to about 11% moisture.

To 100 ml of beer containing 0.65 ppm w/v of diacetyl is added 140 mg of encapsulated enzyme material. After 24 hours at 2°C the diacetyl content is 0.38 ppm w/v.

Example 4: 400 mg of freeze-dried reductase enzyme preparation and 160 mg of reduced nicotinamide adenine dinucleotide are mixed with 50 ml of a 2% w/w aqueous solution of sodium alginate. The mixture is then added dropwise to a vessel, with agitation, containing 400 ml of a 2.5% w/v solution of calcium chloride. The beadlets are filtered and fluid-bed dried to about 85% solids.

ENZYME REACTORS

FORAMINOUS CONTAINERS IN A STIRRER SHAFT

It is usually desirable to use insoluble catalysts in such a way that they offer a large surface area per unit weight of catalyst. For this reason, the water-insoluble carriers chosen for insolubilizing otherwise soluble catalysts are usually in a particulate and/or porous form. The insoluble carriers may be organic such as cellulose or any of various polymers available, or inorganic such as small glass beads or particles of porous glass.

In many respects, porous glass particles provide an ideal carrier for such catalysts as enzymes. For example, porous glass is dimensionally stable and it is relatively inert. Also, it can be easily cleaned or sterilized prior to catalyst attachment. Further, being porous, it offers an extremely large surface area per unit weight (e.g., carriers of powdered porous 96% silica glass of 350 ± 50 A pores of less than 350 mesh are commonly used for insolubilizing enzymes).

Thus, a greater amount of catalyst can be attached to the carrier by utilizing the inner surface area of the pores. By utilizing a porous carrier in comminuted or particulate form, an even greater carrier surface area is provided. Porous catalyst carriers or supports are being used more and more extensively because of the large surface area per unit weight they provide.

However, for almost all solid catalyzed fluid phase reactions, pore diffusion resistance can play an important role in determining the rate of reaction. Therefore, it has been found highly advantageous to utilize porous catalyst supports in very small size particles, thereby reducing the pore length through which reactants must pass to effectively utilize available surface area to which the catalyst is attached. Such particulate, porous catalyst carriers can be used in a variety of chemical reactors. For example, porous catalyst supports may be used in a batch reactor, a continuous stirred tank reactor (CSTR), a fixed bed reactor, and a fluidized bed reactor.

In using the above reactors, however, it has been found the much desired small catalyst support particle size has several disadvantages in its practical applications. Industrial scale utilization of a particulate catalyst support in a fixed bed reactor is, in many cases, impractical due to extremely high resistance to the fluid flow offered by that type of packing.

On the other hand, utilizing a particulate catalyst support in a batch or CSTR reaction, frequently results in attrition of the catalyst particles. The same problems are commonly encountered with fluidized bed reactors. There are yet other problems associated with using particulate catalyst supports in the above reactors. For example, in using catalyst supports or carriers of high surface area there is commonly encountered a film diffusion resistance on the particle surface which hinders catalytic action.

Also, in many catalytic reactions, solid products are formed which settle on the catalyst thereby gradually diminishing available catalytic area. In addition, this makes it difficult to replace or regenerate the catalyst composite, and, in some cases, to recover the product sought.

Lastly, in many catalytic reactions, it is desirable to keep solids separate from the liquid phase, thereby limiting catalytic action to the liquid phase alone. For example, if reactants containing solid materials are utilized in a fixed bed reactor containing a catalyst support, the solid materials tend to clog the reactors. Thus, in view of the numerous disadvantages associated with the use of particulate and/or porous catalyst supports, there has been a long felt need for either a method or apparatus to facilitate the use of such catalyst supports.

N.B. Havewala and H.H. Weetall; U.S. Patent 3,767,535; October 23, 1973; assigned to Corning Glass Works have found that the disadvantages associated with using particulate and/or porous catalyst supports can be overcome with a stirring device that can be used in batch and continuous stirred tank reactors. The device consists of a stirrer shaft that can be driven by conventional means such as by a rotating and/or reciprocating motor or by hand, and one or more foraminous-containers for the particulate catalyst which are attached to the shaft. The foraminous containers also act as impellers when the shaft is rotatably and/or reciprocably driven in a fluid medium.

The containers comprise screen packets for holding the particulate catalyst while permitting inward and outward diffusion of reactants and products. The containers may be detachably mounted on the stirrer shaft and/or have closeable openings to facilitate the replacement or regeneration of the particulate catalyst. By controlling the speed of the shaft in a fluid medium, film diffusion resistance can be easily controlled or at least minimized.

The stirrer can be made from any materials which will not significantly affect the reaction in which they are used. The materials need only be sturdy enough to be driven or agitated in a fluid medium without significant loss of structural integrity for the rotation and/or reciprocating speed chosen for use in a medium of given viscosity. For example, if high speed rotation or reciprocation was anticipated in a relatively viscous reaction medium, it would be best to use materials of known structural stability under adverse conditions, e.g., stainless steel can be used to form the shaft as well as the packets attached to the shaft by means of stainless steel bolts. On the other hand, if a relatively slow agitating speed is used in a medium of low viscosity, the materials used to construct the stirrer could be of any lesser strength material which would not interfere with the reaction, e.g., plastics, glass, wood, and the like. The materials comprising the shaft and parts of the packets need not be the same.

As to the packets, the only requirements are that they be attachable to the shaft in such a way that they function not only as containers for the particulate catalyst, but also as impellers for the stirring device. As used herein, to function as an impeller means that the packets, when attached to the shaft which is rotated, reciprocated, or otherwise moved about in a fluid medium, tend to disrupt the apparent stillness of the fluid.

The shaft can have one or more packets thus attached. More than one packet is attached to the shaft to balance the rotation or other movement of the stirrer shaft in use. Also, it is desirable to use more than one packet to carry greater amounts of the particulate catalyst and thus hasten the reaction, or, in some cases, to carry more than one catalyst.

The stirring assembly may be of any size commensurate with the magnitude of the reaction desired. Thus, the stirrer may be only a few inches in greatest diameter for use in a small beaker, or it may be many feet across for use in a large tank.

The foraminous container for holding the particulate catalyst should have generally evenly distributed openings, small enough to adequately contain the particulate catalyst yet large enough to permit diffusion of the reactants into the container and thereby permit intimate contact with the catalyst. The maximum size of the openings should be smaller than the average particulate catalyst size. Likewise, the products of catalysis should be able to diffuse out of the container and into the reaction medium.

The openings may be in the form of perforations distributed about the container or they may result from using a screen of known mesh size to construct a container on a relatively ridged support structure. Stainless steel screen has been found to be an excellent material from which to construct the containers. Thus, when particulate catalysts such as enzymes insolubilized by bonding to porous glass particles are used, a screening of 400 to 40 mesh NBS has been found suitable since the carrier particles commonly used will generally not pass through a 40 mesh screen.

It has been found that generally the particulate catalyst used in this process should be at least 70 mesh size to minimize or eliminate the problems referred to above.

Figure 4.1 shows a side elevational view of a stirrer. In Figure 4.1, there can be seen that the stirrer 5 consists of a cylindrical shaft 7 about which is frictionally attached a flanged disc member 8. The flanged disc member 8 may also be secured about the shaft 7 by means of one or more set screws (not shown) which pass through the collar portion 9 to engage the shaft 7.

FIGURE 4.1: FORAMINOUS CONTAINER ON A STIRRER SHAFT

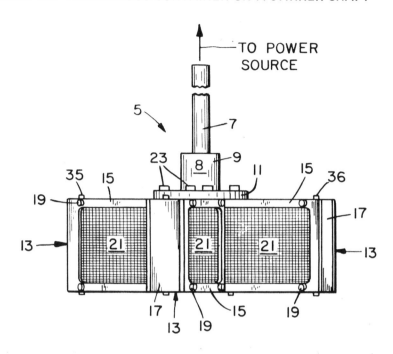

Source: N.B. Havewala and H.H. Weetall; U.S. Patent 3,767,535; Oct. 23, 1973

The flanged portion **11** of the flanged disc member **8** is securely positioned on the shaft **7** distal to the power source not shown. Radially attached to the disc portion **11** of the flanged disc member **8** are box-like foraminous packets **13**, three of which are shown. As can be generally seen, each of the foraminous packets **13** shown in Figure 4.1 have a box-like shape having for two sides meshed screen members **21** held in place against a frame member (not shown in the figure) by retaining members **15** secured to the frame members by four screws **19** on each screened side of the container.

BIOCATALYTIC MODULE CONTAINING ENZYME-MEMBRANE COMPLEX

W.R. Vieth, S.S. Wang, S.G. Gilbert and K. Venkatasubramanian; U.S. Patent 3,809,613; May 7, 1974; assigned to Research Corporation have described a biocatalytic module comprising an immobilized enzyme-membrane complex formed into a film having an elongate surface for enzyme-substrate contact; support means maintaining the elongate surface in a predetermined, substantially elongate form; spacing means substantially preventing different portions of the elongate surface from touching each other; and conduit means minimizing dispersion of a fluid substrate by substantially restricting the flow of the substrate to a direction parallel to the axis of the elongate surface. The biocatalytic modules are easily formed into insertable modules having high surface-to-volume ratios, and are useful for a wide variety of applications in enzymology.

The process is described with reference to Figure 4.2. Figure 4.2a is a perspective view, partially in cross section, of one preferred embodiment. Figure 4.2b is a partial cross section view taken along line 2—2 of Figure 4.2a. Figure 4.2c is a graph of enzyme activity plotted as a function of the volume of washings of the module shown in Figures 4.2a and 4.2b. Figure 4.2d is a schematic view, partially in cross section, of a biocatalytic module formed from a plurality of rods individually layered with an immobilized enzyme-membrane complex. Figure 4.2e is a cross-sectional view taken through line 5—5' of Figure 4.2d. Figure 4.2f is a schematic view, partially in cross section, of a biocatalytic module formed from a helical support layered with an immobilized enzyme-membrane complex. Figure 4.2g is a similar view of a module having a tubular immobilized enzyme-membrane complex fitted over a helical support.

While any immobilized enzyme capable of existing in or on a membrane, film, or sheet may be used to form modules, the description below will be given in terms of using complexes of collagen with various enzymes, since they are simple to prepare and handle. In general, immobilized enzymes capable of existing in or on a membrane, film, or sheet include a wide variety of enzymes complexed with proteins, polypeptides, synthetic polymers and other.

Such films or membranes may contain one or more enzymes (e.g., a combination of amylases). Different modules containing different, individual enzymes or different combinations of enzymes can be staged in series or in parallel arrangement to suit each individual process. In the case of nondiffusible primary substrates, the modules of immobilized enzymes can be used in conjunction with a digestion tank where free enzymes, microencapsulated enzymes, or other catalysts are used for primary substrate degradation.

An ultrafiltration process can be used to recover and recycle the enzymatic catalysts used in the digestion tank. Furthermore, suitable enzyme-membrane complexes themselves can also be used as the ultrafiltration membranes of choice to catalytically facilitate the desired membrane transport processes. The overall reaction-separation system may employ a batch operation, a semibatch operation, a continuous operation, or any combination of these contact modes.

Briefly, preferred enzyme-collagen complexes prepared by swelling a collagen membrane, washing it, and contacting it with an aqueous solution of enzyme thereto for a period of time sufficient to form complexes between the enzyme and the collagen carrier. The membrane may then be poured into sheets, such as by layering over a suitable supporting base, for example, cellulose acetate. Such materials may be dried and stored for long periods of time prior to use. Such films are ideally suited to spiral winding.

Suitable methods for forming a collagen membrane include first treating the collagen source with a proteolytic enzyme solution to dissolve the elastin which encircles and binds the collagen fibers. The collagen source is then washed with water, and the soluble proteins and lipids removed by treatment with a dilute aqueous solution of a chelating agent, such as EDTA. The fibers are then swollen in a suitable acid or base so as to form a collagen fiber dispersion which can then be formed into a suitable membrane by any convenient technique such as by extruding, casting or electrodeposition methods.

While the thickness of the film used to form the modules is not critical except within broad limits, it will be realized that enzymatic activity may be concentrated at the surface thereof. Accordingly, the minimum thickness will be determined by the required mechanical strength and lack of pinholes or other structural defects, depending upon the particular film and/or substrate utilized. The maximum thickness will generally be determined by economic considerations.

Films which are over 0.1 mm thick further tend to result in a lower surface-to-volume ratio, resulting in a loss of efficiency. The collagen membranes useful in this process generally have a dry thickness of from 0.005 to 0.1 mm, preferably from 0.01 to 0.05 mm.

Following the swelling treatment, the swollen collagen membrane is washed thoroughly with water and/or a buffering compound in order to bring the pH level of the membrane within the acceptable range for the particular enzyme being complexed. The swollen, washed membrane is then soaked in an aqueous enzyme solution to allow the enzyme to diffuse into the membrane, usually overnight under refrigeration. The enzyme-collagen complex may be dried, either before or after layering on a supporting substrate such as cellulose acetate film, and may then be stored under refrigeration for a long period of time.

Once the desired immobilized enzyme-membrane complex has been prepared, the biocatalytic modules are prepared by forming the complex into a plurality of substantially parallel capillaric components, or by forming the complex into a strip-wrapped rod/wire bundle, or a strip-wrapped spring, or a tubular reactor. In the spiral and tubular reactors, a unique feature is that the flow of substrate occurs both over and through the membrane.

While intact sheets of the membrane complex are used in tubular and capillaric coil reactors, such sheets are cut into thin strips and wound around metallic rods/wires to form the strip-wrapped rod/wire bundles and the strip-wrapped spring reactor. Higher surface-to-volume ratios are obtained in this way while retaining good mechanical support and contact efficiency. By suitable reduction of the strip dimension, e.g., to small fiber dimensions, it is possible to obtain surface-to-volume ratios of as high as 30,000. Of course, such fibers might be either solid or hollow.

A preferred configuration which is easy to prepare and handle is one which is formed from an axial support spirally wound with an immobilized enzyme-membrane complex, with the overlapping spiral layers formed into a plurality of substantially parallel capillaric components by spacer members oriented substantially parallel to the axial rod used to separate overlapping spiral layers.

Referring to the drawings, Figure 4.2a is a perspective view, partially in cross section, of one preferred embodiment described more particularly below. An axial rod **10** functions as a central supporting member, which is spirally wound with a membranous immobilized enzyme **12**. The overlapping spiral layers are formed into a plurality of parallel capillaric components **14** by spacer rods **16** oriented substantially parallel to the axial rod **10**, and separating the overlapping spiral layers. If desired, the entire module may be provided with an inert protective casing **18** and a filter **20**, in a configuration such as shown in Figure 4.2a to provide an insertable module. Membranous layers **12** may be an immobilized enzyme-membrane complex alone, or such a complex layered on one or both sides of a supporting carrier (not shown).

FIGURE 4.2: BIOCATALYTIC MODULE CONTAINING ENZYME MEMBRANE COMPLEX

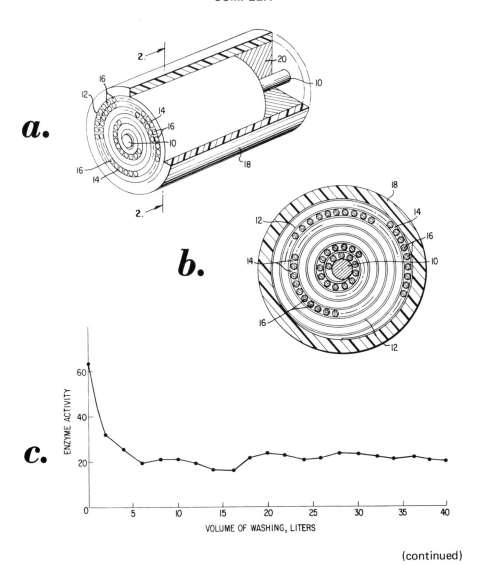

(continued)

FIGURE 4.2: (continued)

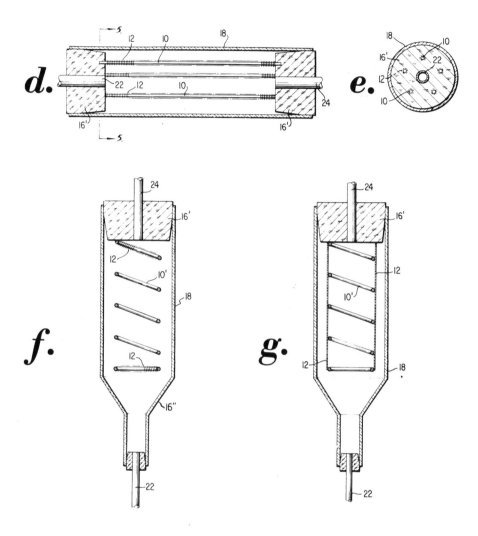

Source: W.R. Vieth, S.S. Wang, S.G. Gilbert and K. Venkatasubramanian; U.S. Patent 3,809,613; May 7, 1974

The axial support and spacer members may be of any material which is inert with respect to the reaction conditions to be encountered and the enzyme-membrane complex being used. When using glass, some loss of activity may be experienced due to dissolution of the glass. Also, since some glass is attacked by alkali, other material may be preferred when the enzyme to be used has an optimum pH on the alkaline side.

The axial support can either be rod-shaped glass, plastic, ceramic, and the like, or a thin sheet or membrane of spongy material such as polyurethane or other plastic foam. In the case of a rod-shaped support, it may be solid or hollow, and preferably will have an outside diameter of from 2 to 7 mm for a module with an overall volume of $\pi(3.13)^2 \times 7.5$ cm^3. Use of larger diameter supports is acceptable, but decreases the surface:volume ratio of the module. Axial supports smaller than indicated are likewise acceptable, but are somewhat more difficult to mount properly.

One end of the immobilized enzyme-membrane complex is affixed to the axial support by any suitable means to facilitate spiral winding. A plurality of spacer members are then affixed to the membrane complex, facilitated by the fact that the film sticks to the glass rods when the rod is wetted with water. Suitable distances between rods vary as the winding procedure progresses.

Preferably, the spacers will be located at intervals small enough to prevent the membrane from coming into contact with overlapping and underlapping layers when wound, and at intervals large enough to maximize the surface-to-volume ratio of the module. The spacer members may be of any desired cross-sectional configuration, although circular or elliptical cross-sectional configurations are preferred. In use, the spirally wound module may shrink over a period of time.

Additional immobilized enzyme-membrane complex material may be added to restore the original volume of the module merely by wrapping additional complex around the spirally wound module. Preferably, the entire module will be encased in a hollow, inert protective casing, and, if desired, a coarse filter element may be inserted into one end thereof to filter substrate solution at the inlet end of the module.

Another preferred configuration which is easy to prepare and handle is one which is formed from a bundle of straight rods wrapped with thin strips of immobilized enzyme-membrane complex as shown in Figures 4.2d and 4.2e. The straight rods used may be any suitable material such as glass, plastic, ceramic, wire, or the like. The length-to-diameter ratio of the rods may vary depending upon the material used. Preferably, this ratio will be small enough to give the material enough strength to prevent or resist deformation, and large enough to maximize the surface-to-volume ratio of the reactor.

The rods are then wrapped roundabout with thin strips of enzyme-membrane complex. The suitable width of the enzyme-membrane complex depends on the diameter of the rod used. For example, a width of 2.5 mm of 1 mil thick membrane complex is suitable for a rod with a diameter of from 1 to 3 mm. Preferably, the rods are then held in parallel position by capping with disk-shaped headers with suitably drilled holes or in the holes of screen type headers similarly placed. Preferably, the entire module will be encased in a hollow biochemically inert protective casing.

Referring to Figures 4.2d and 4.2e, a biocatalytic module is shown formed from a plurality of support means **10** for maintaining the elongate surface of the immobilized enzyme-membrane complex **12** in a predetermined, substantially elongate form. Each of the rods shown is individually layered with an immobilized enzyme-membrane complex **12** formed into a film having an elongate surface for enzyme-substrate contact. Spacing means **16'**, illustrated as disk-shaped header caps with suitably spaced holes for receiving the ends of the wound support rods **10**, serve to prevent different portions of the elongate contact surface of the immobilized enzyme-membrane complex **12** from touching each other.

A biochemically inert, tubular protective casing **18** cylindrically surrounds the parallel support rods **10**, thereby minimizing dispersion of the fluid substrate by substantially restricting its flow to a direction parallel to the axis of the elongate substrate contact surface of immobilized enzyme-membrane complexes **12**. To facilitate use as an insertable module, the biocatalytic module of this example is provided with a substrate inlet **22** and a substrate outlet **24**, such as the hollow glass tubing shown.

If the preferred supporting materials used in making the membrane strip wrapped rod bundle are easily molded into a spring form it is also preferable to use this form, as shown in Figure 4.2f. The immobilized enzyme-membrane complex **12** is formed into a film having an elongate surface for enzyme-substrate contact by wrapping it around a spring-shaped support means **10'**, which maintains the elongate surface in a substantially elongate, helical form.

Support means **10'** may also function for preventing different portions of the elongate surface from touching each other, or spacing means **16'** may be additionally provided (such as in the case of a highly flexible support means) to prevent different portions of the elongate contact surface from touching each other. As shown in Figure 4.2f, spacing means **16** may merely be an insertable cork, or, as shown at **16"** a tapered neck in protective casing **18**. A number of membrane strip-wrapped springs can be nested either axially or concentrically in an inert casing to form a compact biocatalytic cartridge.

Still another preferred configuration is the tubular reactor shown in Figure 4.2g which uses a tube of immobilized enzyme-membrane complex **12** joined from a flat sheet which is then stretched over a spiral tube support means **10'**, such as a spiral of wire, plastic or the like.

A unique feature of the tubular reactor is that the flow of substrate occurs both over and through the tubular membrane. The smaller the diameter of the tube, the larger will be the surface-to-volume ratio. A number of such tubular reactors may be nested in parallel with a suitable mechanically supportive spacer, such as by wrapping them layerwise with a plastic sponge sheet, then inserting this cartridge into an inert casing to form a compact biocatalytic module. In this case, the roles of the membrane and spacer are simply reversed in comparison with the capillaric coil module of Figure 4.2a. The construction of the module shown in Figure 4.2g is similar to that shown in Figure 4.2f and, indeed, may be combined with the construction features shown in Figures 4.2e and 4.2f.

One advantage offered by the biocatalytic modules of this process is that large, molecules, such as antibodies and other proteins, and particulate materials, such as cell suspensions, can readily penetrate the modules, being exposed to the action of the enzyme thereon. This is in marked contrast to prior methods, particularly where the use of ion-exchange materials and molecular sieves results in

entrapment or filtration of materials which are present in the substrate solution
to be treated. The biocatalytic modules do not possess such drawbacks.

Example 1: 6 cc of a 1 mil thick invertase-collagen complex layered on 1 mil
thick cellulose acetate and having overall dimensions of 5.5 cm x 235 cm was
used as membranous enzyme **12** to construct the cartridge shown in Figure 4.2a.
As the central axially oriented rod **10**, a Teflon rod 5.5 cm long and 6 mm in
outside diameter was used. Type 180 glass tubing, 5.5 cm long and having an
outside diameter of 3 mm, was sealed at both ends and used as spacers **16** in
constructing the cartridge.

The invertase-collagen membrane complex **12** is coiled upon the spacers **16**, and
sticks to the glass rods when wet with water. The spacers are located at intervals small enough to prevent the membrane from coming into contact with overlapping and underlapping layers, and at intervals large enough to maximize the
surface-to-volume ratio so as to form a plurality of capillaric components **14**
which present a large surface area for the substrate solution.

In the particular example using a film having a thickness of 1 mil and glass tubes
having an outer diameter of 3 mm, such a spacing is approximately from 6.0 to
13.0 mm for a module with a dimater of 6.25 cm. After coiling the invertase-
collagen membrane complex upon the spacers as indicated above, the cartridge is
then fitted into a tube **18** of any suitable material, such as plastic, to form a
flow-through reactor configuration as indicated in Figure 4.2a. A coarse filter
20 may be affixed to the inlet end when required by the nature of the feed solution.

A series of sucrose hydrolysis experiments was performed in order to test the
biocatalytic cartridge. It was found that after 10 individual experiments were
made, the module shrunk slightly, and it was necessary to add 2 cc of a 1 mil
thick cellulose acetate film to enlarge the shrunken module and restore its snug
fit in the housing. One advantage of the module of this process is that its size
may be varied in this manner.

In testing enzyme activity, 400 cc of a 6% sucrose solution was used as the substrate. The module was connected to a substrate reservoir, and a polarimeter
loop was used to assay enzymatic activity. The flow rate of the substrate solution
through the cartridge was 300 ml/min and the substrate was recycled. The overall reaction time for each experiment was 30 minutes at room temperature.

Before reusing the reactor for another experiment, it was washed by 2 liters of
distilled water. A series of experiments was carried out over a total time span
of seventy days, and the module reactor was stored at 2°C while not being used.
Figure 4.2c shows a plot of reaction rates as percentage conversion of sucrose
against the liters of washings in-between experiments, and indicates that the module shows a stable retention of approximately one-third of its original enzymatic
activity, which remains constant even after 20 liters of washing. The module was
still comparatively active 8 months after these series of experiments was begun,
indicating good stability and the possibilities of long-term use of such a module.

Example 2: A membrane strip-wrapped rod bundle module as shown in Figure

4.2d was prepared. A lysozyme-collagen complex 15 cm x 10 cm and 1 mil thick was slit into thin strips of about 0.25 cm width. These strips were wound around a metallic rod 0.16 cm diameter and 13 cm long. Five such rods were then fitted into a glass tube 1.7 cm diameter and 14 cm long. The rods were held in position by placing them in suitably drilled holes in the corks fitted at the ends of the glass tube. The rods were placed at equal radial distance. The clearance between the glass tube and the periphery of the circle formed by the rods was 0.5 cm.

One-tenth cc of the lysozyme-collagen complex was used. Experiments were conducted to test the performance of the module, using a 300 mg/l suspension of *Micrococcus lysodeikiticus* in a phosphate buffer of pH 7.0, as the substrate. The lysing reaction as indicated by decrease in turbidity was followed spectrophotometrically. Percentage cells lysed in 30 minutes was then calculated and found to be 83 ± 10% (average of two successive experiments).

APPLICATIONS

CHILLPROOFING OF MALT BEVERAGES

B.S. Wildi and D.C. Boyce; U.S. Patent 3,597,219; August 3, 1971; assigned to Monsanto Company have reported improved enzyme processing of fermented malt beverages and other nondistilled alcoholic beverages for purposes of providing a stabilized, enhanced appearance and improved flavor at the time of use. Current enzyme treatments leave active enzyme in the product which may have deleterious effects and which in any event present another active entity, namely, the enzyme, for intake into the human body.

Fermented malt beverages such as beers, ales, and the like are produced by the fermentation with yeast of worts obtained from mashes of barley malt and grains. After fermentation, the beers so obtained are carried through various operations such as cold storage, carbonation, filtration, etc., in order to obtain the clear carbonated beverage ready for packaging. During the brewery operations, the beer is subjected to a process step known in the trade as chillproofing.

When beers are subjected to low temperatures, as occur, for example, during conventional refrigeration and these beers have not been chillproofed, a haze or turbidity forms in the beer as a consequence of the presence of high molecular weight, protein-like compounds and protein complexes involving carbohydrates, phenols, tannins, etc., that tend to become insoluble when the temperature is reduced. Chillproofing is a step in the brewing process that produces a beer which will remain clear and brilliant at low temperatures. The chillproofing process was introduced to brewing when the value of proteolytic enzymes for such use was first demonstrated.

Chillproofing comprises treating the beer or ale after fermentation with certain proteolytic enzymes. During the next phase, or pasteurization, the enzymatic activity is accelerated to prevent formation of haze-producing complexes. A residual enzymatic activity remains after pasteurization. The enzyme papain is commonly used for chillproofing. A heretofore-unrecognized chillproofing sys-

186

tem resides in the employment of polymer-enzyme products in which selected enzymes are covalently bonded to polymeric organic molecules such as EMA.

Currently the brewery adds a specified amount of standard chillproofing enzymes preparation to a known volume of beer, i.e., 1 lb of chillproofing preparation per 100 or 200 barrels of beer. The cost of this treatment ranges between 2 to 4¢ per barrel.

The insoluble polymer-enzyme product can be advantageously introduced into the brewing process at any of several stages. It can be added to the cooled wort prior to fermentation in which case the objectionable precursors of haze are digested during the fermentation process. In this instance, it is advantageous to use a product cross-linked to a somewhat lesser degree so that, while it remains insoluble, it is nonetheless in a gelatinous form which remains in suspension in the beer during the process of removing the yeast which settles after fermentation. Thus the insoluble yeast is removed by decantation techniques while the enzymatic activity continues through the ruh and storage stages and up to the filtration step at which time the insoluble polymer-enzyme product is removed for reuse. The insoluble polymer-enzyme product may also be added at later stages (e.g., during fermentation or storage) in keeping with traditional manufacturing processes.

A further exemplification involves the pumping of storage beer through a filter-type unit containing the insoluble polymer-enzyme product, this unit being maintained at a temperature consistent with the optimum activity-temperature of the specific polymer-enzyme product or products being used. This effluent beer is then cooled, bottled and, where desired, pasteurized.

In the preparation of a water-insoluble EMA-papain, a solution of 0.5 gram of crystalline papain is suspended in 55 ml of 0.05 M acetate buffer, pH 4.5, until a clear solution is obtained. The papain solution is added with stirring to a cold (0° to 5°C) homogenized suspension of 2.5 grams of EMA-21 copolymer in 250 ml of 0.1 M phosphate buffer at a pH of 7.6. Hexamethylenediamine (1.25 g) is added to cross-link the polymer-enzyme complex as it is formed. The reaction mixture is stirred overnight at 4°C. The insoluble EMA papain derivative is separated by centrifugation and washed thoroughly with 0.1 M phosphate buffer, pH 7.5, and 0.1 M NaCl until the washings are free from enzymatic activity. Finally the product is washed with water to remove the salt and the product lyophylized yielding 3.2 grams of EMA-papain retaining 67% of the esterase activity found in the initial crystalline papain.

Example 1: Use of Insoluble EMA-Papain in Beverage Chillproofing (During Fermentation) — 100 barrels of wort is prepared using 60% malt and 40% corn grits. To the wort at 47°F is added the brewer's yeast and this is allowed to ferment for 24 hours. At this time, 75.7 grams of insoluble EMA-papain polymer is added. This product has 300 tyrosine units per mg.

After fermentation is complete (an additional 96 hours), the beer with the EMA-papain product suspended as a gel is decanted from the settled yeast, stored for 7 days at 3°C, filtered, carbonated, stored for 4 to 5 additional days at low temperature, polished, filtered, bottled and pasteurized. The beer produced by this method has superior clarity, stability, and a decidedly improved taste as compared with beer made in the traditional manner by the addition of soluble papain in the cellar, essentially all enzymes having been removed therefrom by filtration removal of insolubles. The EMA-papain is manually removed from the filter, washed with 1 liter of 0.1 NaOH, filtered, followed by 2 washes with 1 liter of 0.2 M acetic acid and 2 washes with 1 liter of water. This washed, used polymer-enzyme product is then put into another 100 barrel batch of fermenting wort which is treated as described above with the same results.

Example 2: Use of Insoluble EMA-Papain in Beverage Chillproofing (in Ruh Stage with Pasteurization) — To 9.3 kg of filtered carbonated ruh beer in a 5 gal stainless stell carbonated beverage can is added 62 mg of EMA-papain insoluble polymer. The can is sealed and mixed by tipping end to end several times. The CO_2 pressure is raised to 10 lb and the can then stored in a cooler at $0°C$ for 7 days with daily mixing. At this time the beer is pasteurized in the can at $60°C$ for 15 minutes and then immediately cooled. The beer is filtered through a 0.22μ millipore filter, by which filtration the insoluble polymer-enzyme product is removed, and then bottled and again pasteurized. Clarity, stability, and taste is superior to the traditional beer and the product maintains these favorable characteristics 6 to 12 months longer than conventionally chillproofed beer.

A further description of process for chillproofing of beverages, especially malt beverages, using insoluble basic polymer-enzyme products, whereby the activity-pH profile and substrate-binding character of normally inactive or nonoptimally active enzymes can be optimized for performance at the pH range of the beverage has been presented by *B.W. Weinrich, J.H. Johnson, B.S. Wildi and D.C. Boyce; U.S. Patent 3,597,220; August 3, 1971; assigned to Monsanto Company.*

Preparation 1: Partial Dimethylaminopropylamine Imide of EMA — A copolymer of ethylene and maleic anhydride, EMA-21 (30 g) was suspended in xylene (500 ml). The mixture was heated to reflux temperature under a nitrogen atmosphere. Dimethylaminopropylamine (14.58 g) was added and the mixture held at reflux for 4.5 hours. Water, a by-product of imide formation, was separated in a Dean-Stark trap and reaction was judged complete when water was no longer being evolved. The product was precipitated by addition of hexane and dried in a vacuum oven at $105°C$. It contained 8.86% N indicating an imide content of 54.3%.

Preparation 2: Partial Dimethylaminopropylamine Imide of Styrene/Maleic Anhydride Copolymer (Precross-Linked) — A cross-linked copolymer of styrene and maleic anhydride (prepared using 2% divinylbenzene as a cross-linking agent) was converted to a partial basic imide derivative as follows. The copolymer (30 g) was suspended in xylene (500 ml) and the mixture refluxed under nitrogen for 1 hour in apparatus fitted with a Dean-Stark trap for collecting any water present or evolved. The mixture was then cooled and dimethylaminopropylamine (12.12 g) added. The mixture was refluxed until no further water was collected (2.1 ml was evolved; theory 2.12 ml). The product was isolated by pouring the cooled solution into 2 liters of hexane, followed by filtration and a wash with 2 liters of hexane. The yield was 37.5 grams or 93.6% of theory. The product contained 8.12% N, indicating an imide content of 77.5%.

Preparation 3: Water Insoluble Basic EMA-Papain — A solution of 0.5 gram of crystalline papain is suspended in 55 ml of 0.05 M acetate buffer, pH 4.5, until a clear solution is obtained. The papain solution is added with stirring to a cold (0° to 5°C) homogenized suspension of 2.5 grams of basic EMA copolymer prepared in Preparation 1 in 250 ml of 0.1 M phosphate buffer at a pH of 7.6. Hexamethylenediamine (1.25 g) is added to cross-link the polymer-enzyme complex as it is formed.

The reaction mixture is stirred overnight at 4°C. The insoluble basic EMA-papain derivative is separated by centrifugation and washed thoroughly with 0.1 M phosphate buffer, pH 7.5, and 0.1 M NaCl until the washings are free from enzymatic activity. Finally the product is washed with water to remove the salt and the product is lyophilized to yield 3.4 grams of EMA-papain, retaining 62% of the esterease activity found in the initial crystalline papain.

Preparation 4: Insoluble Basic EMA-neutral Protease/Papain — A 50:50 mixture (by weight) of neutral protease (isolated from *B. subtilis,* strain AM) and papain is attached to basic EMA-21 (Preparation 2) resulting in the covalent binding of both enzymes to the insoluble basic polymeric network. The product contains 49% of the original papain activity and 36% of the original neutral protease activity.

Preparation 5: Insoluble Basic SMA-Neutral and Alkaline Protease and Amylase Product — A crude *B. subtilis* AM enzyme mixture containing the specified enzymes (0.8 g) is suspended in cold distilled water (60 ml) and stirred magnetically for 1 hour at 4°C. The resulting mixture is then centrifuged at 8,000 rpm for 10 minutes to remove any suspended and inactive solids. The supernatant is separated and made 0.065 M in calcium ion by the addition of 1 M Ca(OAc)$_2$ and the solution is then stirred for 30 minutes in the cold (4°C). The mixture is then centrifuged at 8,000 rpm for 10 minutes to remove precipitated and inactive solids. To the clarified supernatant there is added, with stirring, cold 0.05 M Veronal buffer, pH 7.8.

While the above solutions are being prepared, SMA (0.1 g) (*B. subtilis* enzymes: basic SMA (Preparation 2) 8:1 w/w) is dissolved in dimethyl sulfoxide (10 ml). This solution is added dropwise to the stirred, cold enzyme solution and the mixture is then stirred overnight at 4°C. The mixture is next centrifuged at 8,000 rpm for 10 minutes and the solid product is collected. The solid adduct is washed using twice its volume of cold, distilled water, with stirring and centrifugation. The adducts are washed in this manner 15 times and the product then isolated by lyophilization. The enzymatic activities of the mixed enzyme product are as follows: Amylase 415,000 u/g; neutral protease 1,093,000 u/g; alkaline protease 685,000 u/g.

Example 1: Use of Insoluble Basic EMA-Papain in Beverage Chillproofing (During Fermentation) — 100 barrels of wort are prepared using 60% malt and 40% corn grits. To the wort at 47°F is added the brewer's yeast and this is allowed to ferment for 24 hours. At this time, 65.7 grams of an insoluble basic EMA-papain polymer (prepared as in Preparation 3) is added. This product has 320 tyrosine units per milligrams. After fermentation is complete (an additional 96 hours); the beer with the basic EMA-papain product suspended as a gel is decanted from the settled yeast, stored for 7 days at 3°C, filtered, carbonated, stored for 4 to 5 additional days at low temperature, polished, filtered, bottled and pasteurized. The beer produced by this method has superior clarity, stability, and a decidedly improved taste as compared with beer made in the traditional manner by the addition of soluble papain in the cellar, essentially all enzymes having been removed therefrom by filtration removal of insolubles. The basic EMA-papain is manually removed from the filter, washed with 1 liter of 0.1 M NaOH, filtered, followed by 2 washes with 1 liter of 0.2 M acetic acid and 2 washes with 1 liter of water.

Example 2: Use of Insoluble Basic SMA-B. Subtilis Enzyme Mixture in Beverage Chillproofing (Storage Beer—Heat Exchanger Flow-Through Treatment) — 200 barrels of chilled storage beer is passed through a heat exchange column packed with 3 lb of insolubilized basic SMA-*B. subtilis* enzyme mixture. This column is maintained at 50°C and at a flow rate of 4 gal/min. The effluent temperature is held at that of the influent by supplemental cooling oils. The beer is then filtered, bottled and pasteurized according to traditional methods. Clarity, stability, and taste of the beer is markedly improved and the bottled beverage has exceptional shelf life especially with regard to clarity and taste. The product is an essentially enzyme-free beer which is characterized by favorable and advantageous stable properties of taste and exceptional clarity.

ORAL HYGIENE COMPOSITIONS

Although enzymes have been employed in oral hygiene compositions with some measure of success, the inherent instability of the enzymes themselves, especially in aqueous environments such as characterize may oral compositions, detracts greatly from their effectiveness and from the duration of their activity. Such compositions characterized by inherent instability, which reduces their shelf life greatly, and introduces odor and indefiniteness as to degree of activity remaining at time of use, obviously leave much to be desired in such enzymatically-active oral hygiene compositions. Such problems of inadequate stability and shelf life are at a maximum when combinations of enzymes are employed in oral hygiene compositions, for which reasons combinations of enzymes in oral hygiene compositions have been avoided.

In addition, it should be mentioned that, due to the relatively neutral pH range of the oral cavity, that is, between about 5 and 9, and preferably between about 6 and 8, certain enzymes having their activity or optimum activity outside of the pH range of the oral cavity have been excluded from effective use in such oral hygiene compositions. If they have been present, they have not been active or optimally active at the pH range of the oral activity.

New and improved enzymatically-active oral hygiene compositions would be highly desirable especially such as would have improved substantivity to the teeth, improved color, stability, lack of odor, and adequate periods of shelf life (particularly when in the form of aqueous compositions). These would allow a combination of enzymatic activities without autogenous denaturation of one enzyme by another, or even of a single enzyme by itself, and would moreover permit the employment of additional enzymatic activities in the oral hygiene composition even though the activity or optimum activity pH range of the native enzyme itself is outside of the relatively neutral pH range of the oral cavity.

The procedures of *B.S. Wildi, T.L. Westman and L. Keay; U.S. Patent 3,751,561; August 7, 1973; assigned to Monsanto Company* involve oral hygiene compotions, comprising as an active component at least one polymer-enzyme product where the enzyme is covalently bound. A plurality of enzymatic activities may be present in the form of a plurality of polymer-enzyme products. Such compositions have the advantage that the enzymatically-active component or components are inherently much more stable by the nature of the polymer-enzyme molecule and further because one enzyme does not digest itself or another enzyme and thereby destroy the activity of the composition, since the various enzyme moieties are involved in different environments. Even a greater improvement is realized when a plurality of enzymes are attached to the same polymer molecule.

Thus, in this manner, are provided oral hygiene compositions containing either a single enzyme-polymer product or a plurality of enzyme-polymer products or a polymer-plural enzyme product, or combinations thereof, all of which are enzymatically active, which are exceedingly more stable and long acting in use and less susceptible to deterioration while in storage because the enzymatic components thereof are not subject to desctruction by the same or a different enzymatic component of the composition.

For example, if one or more of the enzymes present in an ordinary oral hygiene composition should be a protease, it will attack other protease molecules, dextranase molecules, amylase; an alkaline protease will attack neutral protease molecules, when both are present; and neutral protease molecules may even attack and digest each other. Such autogenous diminution of enzymatic-activity is not a characteristic of the composition of the present procedures.

Moreover, since some enzymes have a pH activity or a pH optimum activity in a range unsuitable for oral use, for example, dextranase has a pH optimum of 4 to 5, they are either inactive or only marginally active when used in oral hygiene, since obviously the compositions employed will not (and can not from the standpoint of being orally pharmacologically acceptable) be made strongly acid or basic merely to accommodate the activity or optimum activity range of an enzymatically-active component.

When present covalently bound in an anionic polymer molecule, the pH optimum activity is generally substantially increased, and when bound into a cationic polymer, the pH optimum activity of the enzyme is geneally substantially decreased, so that oral hygiene compositions having enzymatically-active components which are active or optimally active within the relatively neutral pH ranges of the oral cavity are now possible which were heretofore impossible (due to the fact that the native enzyme was either inactive or only marginally active within such pH range).

For example, EMA-dextranase, one of the polymer-enzyme products which may be employed has its pH optimum activity substantially higher than dextranase itself. In addition, innumerable combinations of enzymatic activities can now be embodied into oral hygiene compositions, in the form of water-insoluble or water-soluble materials, each being enzymatically-active and independently capable of degrading components which attach themselves to surfaces of the oral cavity and invite infection or bacterial infestation resulting in tooth decay and other undesirable oral health problems.

The general procedure employed consisted of allowing cold solutions of enzymes in appropriate buffers to react overnight at 4°C with cold, homogenized polymer, e.g., EMA, suspensions. EMA 21 was preferably employed, which had a molecular weight of about 20,000 to 30,000. Other molecular weight polymers may also be used. For example, EMA 11, having a molcular weight of about 2,000 to 3,000, is preferred for a soluble modified enzyme product, and EMA 31, having a molecular weight of about 60,000, may also be employed.

Separation of soluble and insoluble adducts, after reaction, was achieved by centrifugation in the cold centrifuge, about 10,000 rpm and 10 minutes centrifugation time). The soluble adducts were exhaustively dialyzed against water in the cold and the lyophilized. Insoluble adducts were washed (and centrifuged), usually 10 times with cold buffer and 5 times with cold distilled water and then lyophilized.

Reactants utilized in the production of *Bacillus subtilis* enzyme mixture/EMA-21 adduct were prepared to the following requirements:

(1) Anhydrous EMA-21 was prepared from HEMA-21 (hydrolyzed EMA) by heating in vacuum at a temperature of 105°C overnight. The molecular weight of EMA-21 is approximately 20,000 to 30,000. In order to insure a maximized anhydride content a freshly prepared anhydride polymer is utilized or the water of hydrolysis is separated therefrom in a boiling xylene suspension of the polymer.

(2) The Veronal buffer utilized was 0.05 M with a pH of 7.8.

(3) The calcium acetate solution employed was 1 M. This was added in twice the volume in order to bring the enzyme solution to the required calcium ion concentration. Alternatively, a 2M concentration solution was employed.

Different samples of *Bacillus subtilis* AM enzyme mixture, each containing neutral protease, alkaline protease, and amylase, were employed. The material was of three types as follows:

(1) *B. subtilis* strain AM enzyme mixture with an activity of 1.9×10^6 protease units per gram (pH 7) and partly insoluble.

(2) *B. subtilis* strain AM enzyme mixture with an activity of 1.0×10^6 protease units per gram (pH 7) and partly insoluble.

(3) *B subtilis* strain AM enzyme mixture with an activity of 1.43×10^6 protease units per gram (pH 7) and completely soluble.

The crude *B. subtilis* enzyme mixture is suspended in cold distilled water and stirred magnetically for one hour at 4°C. The resulting mixture is then centrifuged at 8,000 rpm for 10 minutes to remove suspended and inactive solids.

(This step is omitted for the completely soluble enzyme system.) The supernatant is separated and made 0.065 M in calcium ion by the addition of 1 M Ca(OAc)$_2$ and the solution is then stirred for 30 minutes in the cold (4°C). The mixture is then centrifuged at 8,000 rpm for 10 minutes to remove precipitated and inactive solids.

To the clarified supernatant there is added, with stirring, cold 0.05 M Veronal buffer, pH 7.8. While the above solutions are being prepared an appropriate quality of EMA (*B. subtilis* enzymes: EMA 21, 8:1 w/w) is dissolved in dimethyl sulfoxide. This solution is added dropwise to the stirred, cold enzyme solution and the mixture is then stirred overnight at 4°C. The mixture is then centrifuged at 8,000 rpm for 10 minutes and the solid product is collected. The solid adduct is washed using twice its volume of cold, distilled water, with stirring and centrifugation. The adducts are washed in this manner 15 times and the product was then isolated by lyophilization.

The yield of insoluble products is advantageously achieved, when desired, by performing the reaction in the presence of a crosslinking agent such as hexamethylenediamine, e.g., at a 1 to 2% concentration relative to the amount of polymer employed. The following example illustrates the process.

Example: Typical mouthwash compositions are prepared according to the following general specifications:

	Weight %
Water	30 – 99
Alcohol	0 – 70
Glycerol	0 – 25
Surfactant	0.1
Antimicrobial agent, e.g., cetylpyridinium chloride	0.01 – 0.1
Flavoring	0.1 – 0.2
Coloring	0 – 0.2
Polymer-enzyme product	0.1 – 2

When one or more, preferably a plurality of polymer-enzyme products are used as enzymatically-active ingredients in the formulation, the compositions are stable even after long periods of storage, long-acting in use by virtue of their stability and substantivity, and extremely effective in removing stains from teeth as well as retarding soft accretions and calculus if used over an extended period. The compositions are most acceptable in appearance and also most effective when they embody water-soluble polymer-enzyme products but longer-acting when they embody the insoluble products.

The polymer-plural enzyme products appear to be most effective, as are those embodying a polymer-neutral protease or a polymer neutral protease plus a polymer-dextranase, and especially a polymer-neutral protease/dextranase product.

COMPANY INDEX

The company names listed below are given exactly as they appear in the patents, despite name changes, mergers and acquisitions which have, at times, resulted in the revision of a company name.

INVENTOR INDEX

U.S. PATENT NUMBER INDEX

NOTICE

Nothing contained in this Review shall be construed to constitute a permission or recommendation to practice any invention covered by any patent without a license from the patent owners. Further, neither the author nor the publisher assumes any liability with respect to the use of, or for damages resulting from the use of, any information, apparatus, method or process described in this Review.

MICROBIAL ENZYME
PRODUCTION 1974

by Sidney J. Gutcho

Chemical Technology Review No. 28

Enzymes are catalysts of biochemical origin and, unlike metal catalysts, are capable of extraordinary specificity and reactivity in biological and chemical systems.

Enzymes produced by microbes are being used in a large number of industries. The recognition of their commercial applications and advantages has given their production an added impetus. Not the least is the fact that they do not impart any toxicity to their substrates under controlled conditions. Also, they can be inactivated readily by heat or a change in pH, when their services are no longer wanted, and usually need not be removed from the material in which they have produced a chemical reaction. The crystallization of enzymes has led to their identification as proteins.

The production of enzymes has become very important and the search for new biological sources is going on every day. Since microbial cultures can be controlled closely and are independent of seasons, and their nutrients can be varied at will, the production of microbial enzymes has a bright future. Such enzymes are used as pharmaceuticals, in the manufacture of pharmaceuticals, in detergents, as additives to foods, in the manufacture of a variety of food products, as diagnostic reagents, as reagents for the production of chemicals, and in the treatment of industrial wastes.

The production of microbial enzymes, as described in this book, is proceeding at an ever-increasing pace. This book is written for microbiologists, enzymologists, biochemists, food technologists, and others who are interested in the economical production and efficient application of microbial enzymes.

A partial and condensed table of contents follows. Numbers in parentheses indicate a plurality of processes per topic. Chapter headings and some of the more important subtitles are given.

ISBN 0-8155-0532-9

272 pages